ADVANCED LEARNER'S DICTIONARY *Of* MATHEMATICS

ADVANCED LEARNER'S DICTIONARY
Of
MATHEMATICS

Compiled & Edited by
A Team of Experts

ANMOL PUBLICATIONS PVT. LTD.
NEW DELHI - 110 002 (INDIA)

ANMOL PUBLICATIONS PVT. LTD.
4374/4B, Ansari Road, Daryaganj
New Delhi - 110 002

Advanced Learner's Dictionary of Mathematics

First Edition, 2000

ISBN 81-261-0460-0

PRINTED IN INDIA

Published by J.L. Kumar for Anmol Publications Pvt. Ltd., New Delhi and Printed at Arora Offset Press Laxmi Nagar, Delhi 110 092.

Preface

Mathematics is a group of related sciences including algebra, geometry and calculus, concerned with the study of number, quantity, shape and space and their interrelationship by using a specialized notation. Mathematical operations and processes involved in the solution of a problem or study of some scientific field. Conventionally considered a rough and difficult subject, mathematics is really very interesting and systematic subject having many curiosities and probabilities for finding solutions.

This *Advanced Learners' Dictionary of Mathematics* is basically designed for students and teachers as well. It is the result of a careful analysis of the needs of advanced level students and should be an invaluable aid in their day-to-day learning.

As it is extremely difficult to limit the contents of such a dictionary at 'Advanced' level standard, many headwords are included that are above this level and that may be of interest to Advanced Learners. A few headwords that are not strictly scientific are also included because of their common usage in everyday scientific terminology. It has been our aim to provide explanations that are easily understood, whilst maintaining academic standards.

We have tried to include the more important terms used at advanced level, but would welcome any comments regarding serious omissions or indeed about the content of any of the entries.

We would like to thank all the people who have cooperated in producing this dictionary. We are also grateful to many people who have given additional help and advice.

—Editors

Preface

Mathematics is a group of related sciences, including algebra, geometry and calculus, concerned with the study of number, quantity, shape and space and their interrelationship by using a specialised notation. Mathematical operations and processes involved in the solution of a problem or field or some scientific field. Conventionally considered a rough and difficult subject, mathematics is really very interesting and systematic subject having many curiosities and probabilities for finding solutions.

This *Advanced Learners' Dictionary of Mathematics* is basically designed for students and teachers as well. It is the result of a careful analysis of the needs of advanced level students and should be an invaluable aid in their day-to-day learning.

As it is extremely difficult to limit the contents of such a dictionary at 'Advanced' level standard, many headwords are included that are above this level and that may be of interest to Advanced Learners. A few headwords that are not strictly scientific are also included because of their common usage in everyday scientific terminology. It has been our aim to provide explanations that are easily understood, whilst maintaining academic standards.

We have tried to include the more important terms used at advanced level, but would welcome any comments regarding serious omissions or indeed about the content of any of the entries.

We would like to thank all the people who have cooperated in producing this dictionary. We are also grateful to many people who have given additional help and advice.

— Editor

A

Abelian group: (commutative group) A type of group in which the elements can also be related to each other in pairs by a commutative operation. For example, if the operation is multiplication, and the elements are rational numbers, then the set is an Abelian group because for any two elements a and b, $a \times b = b \times a$, and all three numbers, a, b, and $a \times b$ are elements in the set. All cyclic groups are Abelian groups.

abscissa: The horizontal or x-coordinate in a two dimensional rectangular Cartesian coordinate system.

absolute: A number or measurement that does not depend on a standard reference value. For example, absolute density is measured in kilograms per cubic metre, but relative density is the ratio of density to that of a standard density.

absolute convergence: The convergence of the sum of the absolute values of term is in a series of positive and negative terms, For example, the series: $1- (1/2)^2 + (1/3)^3 -(1/4)^4 +$.......is absolutely convergent because $1 + (1/2)^2 + (1/3)^3 + (1/4)^4$............is also convergent. A series that is convergent but has a divergent series of absolute values is conditionally convergent, For example,

$1-(1/2)^2+(1/3)^3-(1/4)^4 +$..............

is conditionally convergent because

$1 + (1/2)^2 + (1/3)^3 + (1/4)^4 +$ .... is divergent.

absolute error: The difference between the measured value of a quantity and its true value. Compare relative error.

absolute value: The modulus of a real number, or of a complex number. For example, the absolute value of -2.3, written ½-2.3½ is 2.3. The absolute value of a complex number is also the modulus, for example the absolute value of 2 + 3i is Ö$(2^2 + 3^2)$.

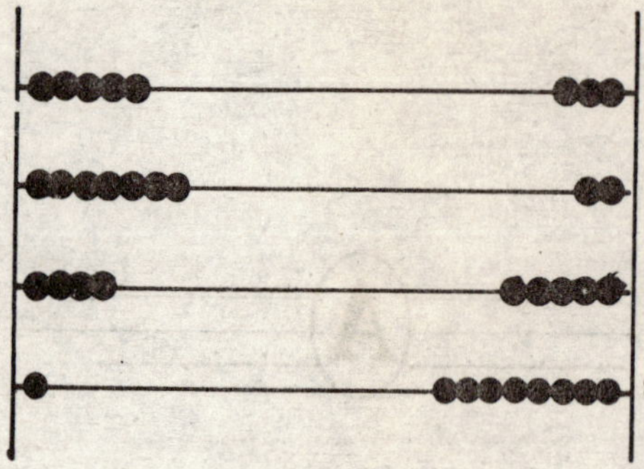

An ababcus showing the number 3258 on the right hand side.

abstract number: A number that does not refer to any specific quantity. For example, the number three when it does not refer to three objects, quantities, etc., but simply the abstract concept of 'three'.

acceleration: Symbol: a The rate of change of speed or velocity with respect to time. The SI unit is the metre per second per second (ms^{-2}). A body moving in a straight fine with increasing speed has a positive acceleration. A body moving in a curved path with uniform speed also has an acceleration, since the velocity is changing. In the case of motion in a circle the acceleration is v^2/r directed to the centre of the circle (radius r).

For constant acceleration:

$$a = (V_2 - V_1)/t$$

V_1 is the speed or velocity when timing starts; V_2 is the speed or velocity after time t.

The equation above gives the mean acceleration over the time interval t. If the acceleration is not constant

$$a = dv/dt, \text{ or } d^2x/dt^2.$$

acceleration of free fall: (acceleration due to gravity) Symbol: g The constant acceleration of a mass falling freely (without friction) in the Earth's gravitational field. The acceleration is towards the surface of the Earth g is a measure of gravitational field strength the force on unit mass. The force of a mass m is its weight W, where W=mg.

The value of g varies with distance from the Earth's surface. Near the surface it is just under 10 metres per second. It varies with latitude, in part because the Earth is not perfectly spherical.

accuracy: The number of significant figures in a number representing a measurement or value of a quantity. If a length is written as 2.314 metres, then it is normally assumed that of the four figures are meaningful, and that the length has been measured to the nearest rnillimetre. It is incorrect to write a number to a precision of, for example, four significant figures when the accuracy of the value is only three significant figures, unless the error in the estimate is indicated.

acre: A unit of area equal to 4840 square yards. It is equivalent to 0.404 68 hectare.

action at a distance: An effect in which one body affects another through space with no apparent contact or transfer between them.

actuary An expert in statistics who calculates insurance risks and relates them to the premiums to be charged.

acute: Denoting an angle that is less than a right angle, that is, less than 90° (or p/2 radian).

addend: One of the numbers added together in a sum.

addition Symbol +: The operation of finding the sum of two or more quantities. In arithmetic, the addition of numbers is commutative (4 + 5 = 5 + 4), associative 2 + (3 + 4) = (2 + 3) + 4, and the identity clement is zero (5 + 0 =5). The inverse operation to addition is subtraction. In vector addition, the direction of the two vectors affects the sum. Two vectors are added by placing them head-to-tail to form two sides of a triangle. The length and direction of the third side is the vector sum. Matrix addition can only be carried out between matrices with the same number of rows and columns, and the sum has the same dimensions. The elements in corresponding positions in each matrix are added arithmetically.

addition formulae: Equations that express trigonometric functions of the sum or difference, off two angles in terms of separate functions of the angles:

$$\sin(x+y) = \sin x \cos y + \cos x \sin y$$

$$\sin(x-y) = \sin x \cos y - \cos x \sin y$$

$$\cos(x+y) = \cos x \cos y - \sin x \sin y$$

$$\cos(x-y) = \cos x \cos y + \sin x \sin y$$

$$\tan(x+y) = (\tan x + \tan y)/(l \tan x \tan y)$$

$$\tan(x - y) = (\tan x - \tan y)/(1 + \tan x \tan y)$$

They are used to simplify trigonometric expressions, for example, in solving an equation. From the addition formulae the following formulae can be derived:

The double-angle formulae.

$$\sin 2x = 2 \sin x \cos x$$

$$\text{Cos}2x = \cos^2 x - \sin^2 X$$

$$\tan(2x) = 2\tan x/(1 - \tan^2 x)$$

The half-angle formulae:

$$\sin(x/2) = \text{Ö}[(1-\cos x)/2)]$$

$$\cos(x/2) = \text{Ö}\sin x/[(1 + \cos x)/2]$$

$$\tan(x/2) = \sin x/(1 + \cos x) = (1-\cos x)/\sin x$$

The product formulae

$$\sin x \cos y = ½[\sin(x+y) + \sin(x-y)]$$

$$\cos x \sin y = ½[\sin(x+y) - \sin(x-y)]$$

$$\cos x \cos y = ½[\cos(x+y) + \cos(x-y)]$$

$$\sin x \sin y = ½[\cos(x-y) - \cos(x+ y)]$$

adjacent: 1. Denoting one the sides forming a given angle in a triangle. In a right-angled triangle it is the side between the given angle and the right angle. In trigonometry the, ratios *et ibis* adjacent side to, the other side lengths arc used to define the cosine and tangent functions of the angle.

2. Denoting two sides of a polygon that share a common vertex.

3. Denoting two angles that share a common vertex and a common side.

4. Denoting two faces of a polyhedron that share a common edge.

algebra: The branch of mathematics in which symbols arc used to represent numbers or variables in arithmetical operations. For example, the relationship:

$$3 ´ (4 + 2) = (3 ´ 4) + (3 ´ 2)$$

belongs to arithmetic. It applies only to this particular set of numbers, On the other hand the equation:

$$x(y + z) xy + xz$$

is an expression in algebra. It is true for any three numbers denoted by x, y and z. The above is a statement of the distributive law of arithmetic: similar statements can be written for that associative and commutative laws.

Much of elementary algebra consists of methods of manipulating equations to put them in a more convenient form, For example, the equation:

$$x + 3y = 15$$

can be changed by subtracting 3y frorn both sides of the equation, giving:

$$x + 3y -3y = 15- 3y$$

$$\text{i.e, } x= 15-3y$$

The effect is that of moving a (+3y) term from one side of the equation to the other and changing the sign. Similarly a multipiication on one of the equation becomes a division when the term is moved to the other side.

$$xy =5$$

becomes

$$x=5/y$$

Ordinary algebra is a generalisation of arithmetic. Other forms of higher algebra also exist, concerned with mathematical entities other than numbers. For example, matrix algebra is concerned with the relations between matrices; vector algebra with vectors; Boolean algebra is applicable to logical propositions and to sets; etc. An algebra consists of a number of mathematical entities (eg. matrices or sets) and operations (e.g. addition or set inclusion) with formal rules for the relationships, between the mathematical entities, Such a system is called an algebraic structure.

algorithm: A mechanical procedure for performing a given calculation or solving a problem in a series of steps. One example is the common method of long division in steps. Another is the Euclidean algorithm for finding the highest common factor of two positive integers.

alternate angles: A pair of equal angles formed by two parallel lines and a third line crossing both. For example, the two acute angles in the letter Z are alternate angles.

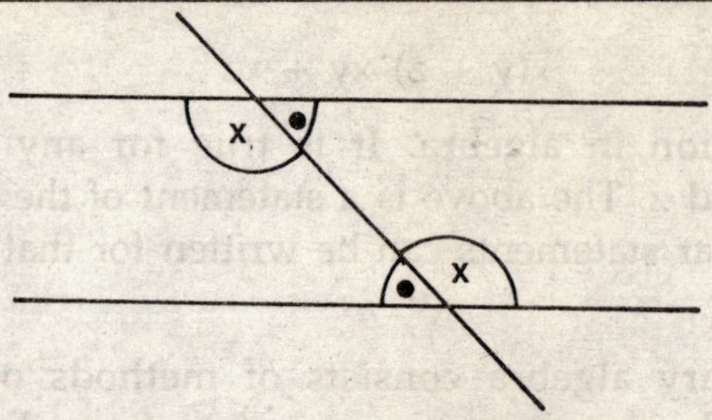

Alternate angles formed by
a line cutting two parallel lines.

alternating series: A series in which the terms are alternately positive and negative, for example:

$$S_n = -1 + 1/2 - 1/3 + 1/4 - \ldots (-1)^n/n$$

Such a series is convergent if the absolute value of each term is less than the preeeditig. The example above is a convergent series.

altitude: The perpendicular distance from the base of a figure to the vertex opposite.

ambiguous: Having more than one possible meaning, value, or solution. For example, an ambiguous case occurs in finding the sides and angles of a triangle when two sides and an angle other than the included angle are know. One solution is an acute-angled triangle and the other is an obtuse-angled triangle.

ampere: Symbol: A The SI base unit of electric current, defined as the constant current that, maintained in two straight parallel infinite, conductors of negligible circular cross section placed one metre apart in vacuum. would produce a force between ihe conductors of 2 ´ 10^{-7} newton per metre.

amplitude: The Maximum value of a varying quantity from it, mean or base value. In the case of a simple harmonic motion—a wave or vibration—it is half the maximum peak-to-peak value.

analog computer: A type of computer in which numerical information is represented in the form of a quantity, usually a voltage, that can vary continuously. This varying quantity is an analogue of the actual data, i.e, it varies in the same manner as the data, but is easier to manipulate in the mathematical operations performed by the analog computer. The data is obtained from some process, experiment, etc.; it could be the changing temperature or pressure in a system or the varying speed of flow of a liquid. There may be several sets of data, each represented by a varying voltage.

The data is converted into its voltage analogue or analogues and calculations and other sorts of mathematical operations, especially the solution of differential equations, can then be performed on the voltage(s) (and hence or, the data they represent). This is done by the user selecting a group of electronic devices in the computer to which the voltage(s) are to be applied. These devices rapidly add voltages, and multiply them, integrate them, etc., as required. The resulting voltage is proportional to the result of the operation. It can be fed to a recording device to produce a graph or some other form of permanent record. Alternatively it can be used to control the process that produces the data entering the computer.

Analog computers operate in real time and are used, for example, in the automatic control of certain industrial Processes and in a variety of scientific experiments. They can perform much more complicated mathematics than digital computers but are less accurate and are less flexible in the kind of things they can do

analogy: A general similarity between two problems or methods. Analogy is used to, indicate the results of one problem front the known results of the other.

analysis The branch of mathematics that uses the concept of limits.

analytical geometry: (coordinate geometry) The use of coordinate systems and algebraic methods in geometry. In a plane Cartesian coordinate system a point is represented by a set of numbers and a curve is an equation for a set of points. The geometric properties of curves and figures can thus be investigated by algebra.

angle: (plane angle) The spatial relationship between two straight lines. If two lines art parallel, the angle between them L; zero. Angles are measured in degrees or, alternatively, in radians. A complete revolution is 360 degrees (360'). A straight fine forms an angle of 180° and a right angle is 90°.

The angle between a line and a plane is the angle between the line and its orthogonal projection on the plane.

The angle between two planes is the angle between lines drawn perpendicular to the common edge from a point one line in each plane. The angle

angular accelleration: Symbol: a The rotational acceleration of an object about an axis; i.e. the rate of change of angular velocity with time:

$$a = dw/dt$$

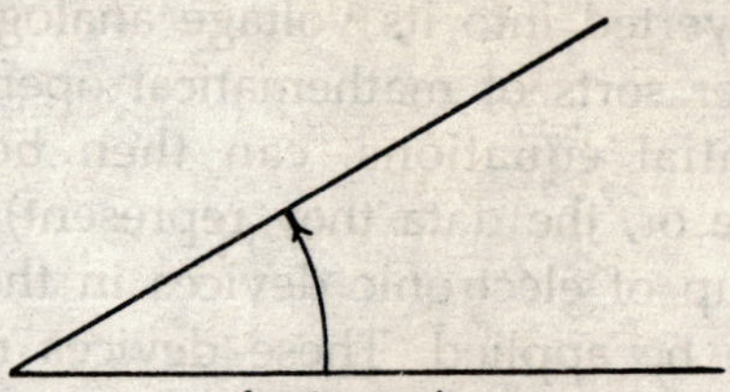

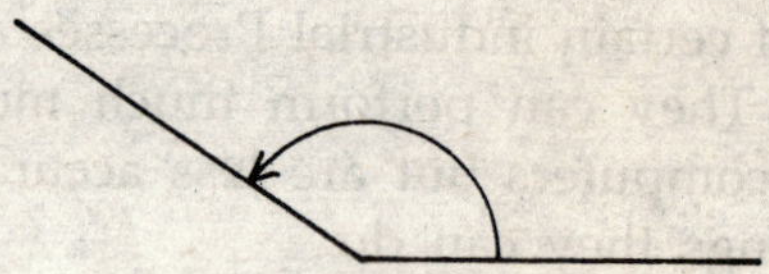

Types of angles

or

$$a = d^2q/dt^2$$

where w is angular velocity and q is angular displacement. Angular acceleration is analogous to linear acceleration.

angular displacement: Symbol :q The rotational displacement of an object about an axis. If the object (or a point on it) moves from point P_1 to point P_2 in a plane perpendicular to the axis, q is the angle PlOP2, where O is the point at which the perpendicular plane meets the axis.

angular frequency: (pulsatance) Symbol: w The number of complete rotations per unit time. Angular frequency is often used to describe vibrations. Thus, a simple harmonic motion of frequency *f* can be represented by a point moving in a circular path at constant speed. The foot of a perpendicular frorn the point to, a diameter of the circle moves with simple harmonic motion. The angular frequency of ibis motion is equal to 2pf, where *f* is the frequency of the simple harmonic motion. The unit of angular frequency, like frequency, is the hertz.

angular momentum: Symbol: L The product of the moment of inertia of a body and its angular velocity. Angular Momentum is analogous to linear momentum, moment of inertia being rite equivalent of mass for rotational motion.

angular velocity: Symbol: w The rate of change of angular displacement with time: w = dq/dt.

anharmonic osillator: A System whose vibration, while still periodic, cannot be described in terms of simple harmonic motions. In such cases, the period of oscillation not independent et the amplitude.

annuity: A pension in which an insurance company pays the annuitant fixed regular sums of money in return for sums of money in turn. An annuity certain is paid for a fixed number of years as opposed to an annuity that is payable only while the annuitant is alive.

annulus: The region between two concentric circles. The area of an annulus is 7(RI r), where R is the radius of the larger circle and r is the radius of the smaller,

antecedent: In logic, the first part of a conditional statement; a proposition or statement that is said to imply another, For example, in the statement 'if it i, raining then the streets are wet', 'it is raining' is the antecedent. Compare consequent. Sec also implication.

anticlockwise: Rotating in the opposite sense to the hands of a clock.

antilogarithm: (antilog) The inverse function of a logarithm, in common logarithms, the antilogrithm of x is 10. In natural logarithms, the antilogarithm of x is e^x.

antinode: A point of maximum vibration in a stationary wave pattern.

antiParallel: Parallel but acting in opposite directions.

apex: The point at the top of a solid, such as a pyramid, or of a plane figure, such as a triangle.

apothem: (short radius) A line segment from the centre of a regular polygon perpendicular to the centre of a side.

applied mathematics: The study of the mathematical techniques that are used to solve the problems. Strictly speaking it is the application of mathematics to any 'real' system. For instance, pure geomatry is the study of entities—lines, point, angles etc.— based on certain axioms. The use of Euclidian geometry in surveying or architecture, navigation or science is applied geometry. The term applied mathematics is used especially for mechanics the study of forces and motion.

Afflonius theorem: The equation that relates the length of a median in a triangle to the lengths of its sides, If a is the length of one side and b is the length of another, and the third side is divided into two equal lengths c by a me" of length m, then

$$a^2 + b^2 = 2m^2 + 2c^2$$

arc: A part of a continuous curve. If the circumference of a circle is divided into two unequal parts, the smaller is known as the minor arc and the larger is known as the major arc.

arc cosecant: (arc cosec, arc sec) An inverse cosecant.

arc cosech: An inverse hyperbolic cosecant.

arc cosh: An inverse hyperbolic cosine.

arc cosine: (arc cos) An inverse cosine.

arc cotangent: (arc cot) An inverse cotangent.

arc coth: An inverse hyperbolic cotangent.

Archimedes' principle: The upward force on an object totally or partly submerged in a fluid is equal to the weight cif fluid displaced by the object.

The upward force, often called the upthrust results from the fact that the pressure in a fluid increases with depth. If the object displaces a volume V of fluid of density r then:

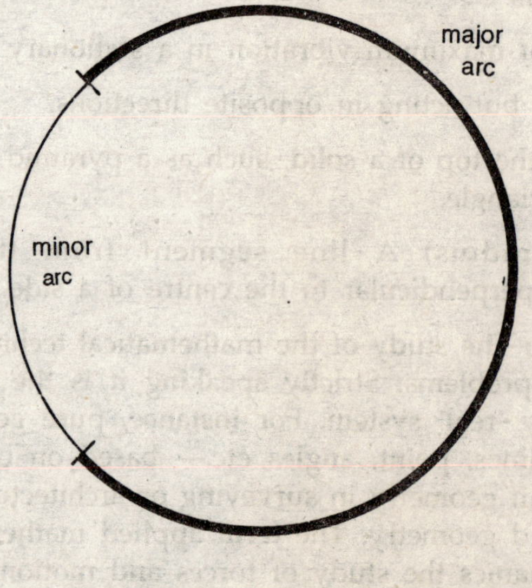

Arc: major and minor arcs of a circle

upthrust u= Vrg

where g is the acceleration of free fall. If the upthrust on the object equals the object's weight, the object will float trigono-metric functions.

arc secant: (arc sec) An inverse secant.

arc sech: An inverse hyperbolic secant.

arc sine: (arc sin) An inverse sine.

arc sinh: An inverse hyperbolic sine.

arc tangent: (arc tan) An inverse tangent.

arc tanh: An inverse hyperbolic tangent.

are: A metric unit of area equal to 100 square metres. It is equivalent to 119.60 sq yd.

area: Symbol: A The extent of a plane figure or a surface, measured in units of length squared. The SI unit of area is the square metre (m^2). The area of a rectangle is the product of its length and breadth. A more complex-shaped area can be estimated by superimposing a square grid and counting the number of whole and part squares that it covers. Formulae for areas can be found integral calculus.

argument: (amplitude) In a complex number written in the form r(cosq + i sinq), the angle q is the argument, It is therefore the angle that the vector representing the complex number makes with the horizontal axis in an Argand diagram.

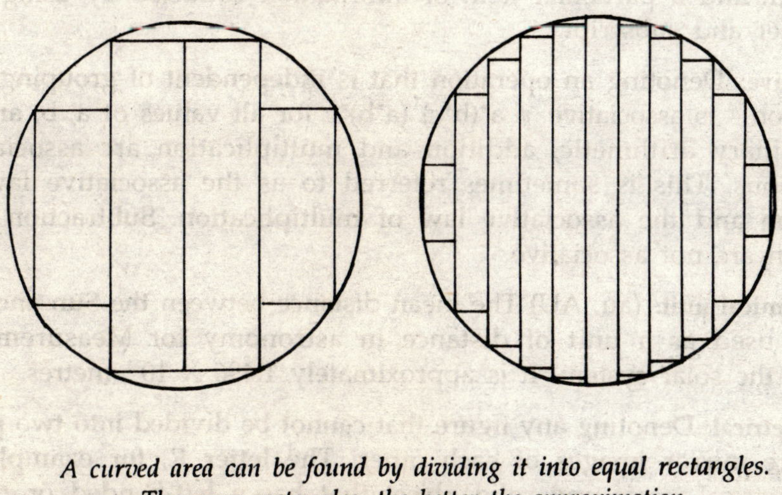

A curved area can be found by dividing it into equal rectangles. The more rectangles, the better the approximation.

arithmetic: The study of the skills necessary to manipulate numbers in order to solve problems containing numerical information. It also involves an understanding of the structure of the number system and the facility to change numbers from one form to another; for example, the changing of fractions to decimals, and vice versa.

arithmetic sequence: (arithmetic progression) A sequence in which the difference between each term and the one alter it is constant, for example, {9, 11, 13, 15, ... }. The difference between successive terms is called the common difference. The general formula for the nth term of an arithmetic sequence is:

$$n_n = a + (n-1)\ d$$

where a is the first term of the sequence and d is the common difference.

arithmetic series: A series in which the difference between each term and the one alter it is constant, for example, 3 + 7 + 11 + 15 + ... The general formula for an arithmetic series is:

$$S_n = a + (a + d) + (a + 2d) + \ldots\ldots\ldots\ldots + [a + (n - 1)d$$
$$= S\ [a + (n - 1)d]$$

arm: One of the lines forming a given angle.

array: An ordered arrangement of numbers or other items of information, such as those in a list or table. In computing, an array has its own name, or identifier, and each member of the array is identified by a subscript used with the identifier. An array can b, examined by a program and a particular item of information extracted by using this identifier and subscript.

associative: Denoting an operation that is independent of grouping. An operation * is associative *if* a*(b*c) (a*b)*c for all values of a, b, and c. In ordinary arithmetic, addition and multiplication arc associative operations. This is sometimes referred to as the associative law of addition and the associative law of multiplication. Subtraction and division arc not associative.

astronomical unit: (au, AU) The mean distance between the Sun and the Earth, used as a unit of distance in astronomy for Measurements within the solar system. It is approximately 1.496×10^{11} metres.

asymmetrical: Denoting any figure that cannot be divided into two parts that are mirror images of each other. The letter R, for example, is asymmetrical, as is any solid object that has a left-handed or right-handed characteristic. Compare symmetrical.

asymptote: A straight line towards which a curve approaches but never meets. A hyperbola, for example, has two asymptotes. In two-dimensional Cartesian coordinates, the curve with the equation $y = 1 / x$ has the lines $x = 0$ and $y = 0$ as asymptotes, since y becomes infinitely small, but never reaches zero, as x increases, and vice versa.

atmosphere: A unit of pressure equal to 760 mmHg. It is equivalent to 101325 newtons per square metre (101325 N m^{-2}).

axial plane: A fixed reference plane in a three-dimensional coordinate system. For example, in rectangular Cartesian coordinates, the planes defined by $x = 0$, $y = 0$, and $z = 0$ me axial planes. The x-coordinate of a point is its perpendicular distance from the plane $x = 0$ and the y and z coordinates are the perpendicular distances from the $y = 0$ and $z = 0$ planes respectively.

axiom: (postulate) In a mathematical or logical system, air initial proposition or statement that is accepted as true without proof and from which further statements, or theorems, can be derived. In a mathematical proof, the axioms are often well-known formulae for *which* the proof has already been established.

axis: 1. A fine about which a figure is symmmetrical. 2 One of the fixed reference fines used in a graph or a coordinate system. 3. A line about *which* a curve or body rotates or revolves.

azimuth: The angle q measured in a horizontal plane front the x-axis in spherical polar coordinates. It is the same as the longitude of a point.

B

ballistics: The study of the motion of objects that are propelled by an external force.

bar: A unit of pressure defined as 10^5 pascals. The millibar (mb) is more common; it is used for measuring atmospheric pressure in meteorology.

bar chart: A graph consisting of bars whose lengths are proportional to quantities in a set of data. It can be used when one axis cannot have a numerical scale, e.g. to show how many pink, red, yellow, and white flowers grow from a Packet of mixed seeds.

barn: Symbol: b A unit of area defined as 10^{-28} square metre. The barn is sometimes used to express the effective cros-sections of atoms or nuclei in the scattering or absorption of particles.

barrel: A US unit of capacity used to measure solids. It is equal to 7056 cubic inches (0. 115 6 ml).

base 1. In geometry, the lower side of a triangle, or other plane figure, or the lower face of a pyramid or other solid. The altitude is measured from the base and at right angles to it.

2. In a number system the number of different symbols used, including zero. For example, in the decimal number system, the base is ten. Ten units, ten tens, etc., are grouped together. In binary numbers, the base is two and the symbols used are 0 and 1, 3. In logarithms, the number that is raised to the power equal to the value of the logarithm. In common logarithms the base is 10; for example, the logarithm to the base 10 of 100 is 2:

$$\mathrm{logt}_{10}100 = 2$$

$$100 = 10^2$$

base unit: A unit that is defined in terms of reproducible physical phenomena or Prototypes, rather than of other units. The metre, for example, is a base unit in the SI, being defined in terms of the wavelength of a particular light emission.

basis vectors: In two dimensions, two non-parallel vectors, scalar multiples of which are added to form any other vector in the same plane, For example, the unit vector i and i in the directions of the x and y-axes of a Cartesian coordinate system are basis vectors, The position vector, OP of the point P(2,3) is equal to 2i + 3j. Similarly, in three dimensions a vector can be written as the sum of multiples of three basis vectors.

batch processing: A method of operation, used especially in large computer systems, in which a number of programs are collected together and input in the computer as a single unit. The programs forming a batch can either be submitted at a central site or at a remote *job* entry site; there can be any number of remote job entry sites, which can be situated at considerable distances from the computer. The programs are then executed as time becomes available in the system, Compare time sharing.

Bayes theorem: A formula expressing the probability of an intersection of two or more sets as a product of the individual probabilities for each. It is used to calculate the probability that a Particular event Bi has occurred when it is known that at least one of the set {B1, B2, ...Bn} has occurred and that another event A has also occurred. This conditional probability is written as P(Bi/A). B1,B2, ...Bn form a partition of the sample space S such that B_1ÈB_2È...ÈB_n = S and B_iÇB_j for all i and j. If the probabilities of B1, B2,.......B_n and all of the conditional probabilities P(A/Bj) are known, then P(Bi/*A)* is given by P(Bi)P(A/ Bi)/P(B_j)P(A/B_j)

Bernoulli trial: An experiment in which there are two possible outcomes, for example, tossing a coin.

Bessel functions: A set of functions denoted by the letter J, that are solutions to Laplace's equation in cylindrical polar coordinates. The solutions form an Infinite series and are listed in tables.

beva: Symbol: B A prefix used in the USA to denote 10^9. It is equivalent to the SI prefix giga.

bias: A property of a statistical sample that makes it unrepresentative of the whole population. For example, if medical data is based on a survey of patients in a hospital, then the sample is a biased estimate of the general population, since healthy people, will be excluded.

biconditional: Symbol: In logic the relationship if and only if that holds between a pair of propositions or statements P and Q only when they are both true or both false. It is also the relationship of logical equivalnce; the truth of P is both a necessary and a sufficient condition for the truth of Q. The truth table definition for the biconditional is shown in the illustration.

binary: Denoting or based on two. A binary number is made up using only two different digits, 0 and 1, instead of ten in the decimal system. Each digit represents a unit, twos, fours, eights, sixteens, etc., instead of units, tens hundreds, etc. For example, 2 is written as 10, 3 is 11, 16 (= 2^4) is 10000. Computers calculate using binary numbers. The digits 1 and 0 correspond to on/off or high/low conditions in an electronic switching circuit.

binary logarithm: A logarithm to, the, base two. The binary logarithm of 2 is written is 1.

binary operation: A mathematical procedure that combines two numbers, quantities, etc., to, give a third. For example, multiplication of two numbers in arithmetic is a binary operation.

binomial: An algebraic expression with two variables in it. For example, 2x + y and 4a + b =0 are binomials.

binomial coefficient: The factor multiplying the variables in a term of a binomial expansion. For example, in $(x + y)^2 = x^2 + 2xy+y2$ the binomial coefficients are 1, 2, and 1 respectively. In general, the rth binomial coefficient in the expansion of (x + y) is n!/[rn(n-r)!]. This is written using the notation ${}_nC_r$. Sec binomial expansion.

binomial distribution: The distribution of the number of successes in an experiment in which there are two possible outcomes i.e. successes and failure. The probability of k successes is

$$b(k, n, p) = n!/k!(n\ k)!'\ p^{n'}q^{n-k}$$

where p is the probability of success and q (=1-p) the probability of failure, on each trial. 'These probabilities are given by the terms in the binomial theorem expansion of $(p + q)^n$. The distribution has a mean np and variance npq. If n is large and p small it can be approximated by a Poisson distribution mean np. If n is large and p is not near 0 or 1, it can be approximated by a normal distribution with mean np and variance npq.

binomial expansion: A rule for the expansion of an expression of the form $(x+ y)^n$. x and y can be any real numbers and n is an integer. The general formula, sometimes called the binomial theorem, is

$$(x + y)^n = x^n + nx^{n-1}y+[n(n-1)/2!]x^{n-2}y^2+...........+y^n$$

When, for example, n= 2, then $(x + y)^n = x^2+2xy + y^2$. If y is smaller than x and is less than one, then when n is very large, the first few terms of the expansion are approximately equal to the equal to the complete series.

bisector: A straight fine or a plane that divides a line, a plane, or an angle into two equal parts.

bistable circuit: An electronic circuit that has two stable states. The circuit will remain in one state until the application of a suitable pulse, which will cause it to assume the other state. Bistable circuits are used extensively in computer equipment for storing data and for counting. They usually have two input terminals to which pulses can be applied. A pulse on one input causes the circuit to change state; it will remain in that suite until a pulse on the other input causes it to assume the alternative state. These circuits are often called flip-flops.

bit: Abbreviation of binary digit, i.e. either of the digits 0 or 1 used in binary notation. Bits are the basic units of information in computing as they can represent the states of a two-valued system. For example, the passage of an electric pulse along a wire could be represented by a 1 while a 0 would mean that no Pulse had passed. Again, the two states of magnetization of, say, the magnetized spots on a magnetic tape can be represented by either a 1 or 0.

bivariate: Containing two variable quantities. A plane vector, for example, is bivariate because it has bath magnitude and direction,

A bivariate tandem variable (X,Y) has the joint probability P(x,y); i.e. the probability that X and Y have the values *x* and y respectively, is equal to P(x) x P(y), when X and Y are independent.

Bliss theorem: The theorem in calculus that relates the definite integral of a Product of two functions to the limit of a series. If f(x) and g(x) are continuous functions in the interval a £ x£ b, and the interval is subdivided into smaller intervals, in the kth subinterval of x, $D_k x$ any two Points x_k and x_h can be selected. The Bliss theorern states that, whether the points x_k and x_i are distinct or coincident:

$$\lim Sf(xk).g(xh)D_k x={}^b\int_a f(x).g(x)dx$$

Board of Trade Unit: (BTU) A unit of energy equivalent to the kilowatt-hour (3.6´10⁶ joules). It was formerly used in the UK for the sale of electricity.

Boolean algebra: A system of mathematical logic that uses symbols and set theory to represent logical operations in a mathematical form, It was the first system of logic using algebraic methods for combining symbols in proofs and deductions. Various systems have been developed and are used in probability theory and computing.

bound: 1. In a set of numbers, a value beyond which there are no members of the set. A lower bound is less than or equal to every number in the set.

boundary condition: In a differential equation, the value of the variables at a certain point, or information about their relationship at a point, that enables the arbitrary constants in the solution to be determined. For example, the equation

$$d^2y/dx^2 - 4dy/dx + 3y = 0$$

has a general solution

$$y = Ae^{-x} + Be^{-3x}$$

where A and B are arbitrary constants. If the boundary conditions are $y = 1$ at $x=0$ and $dy/dx = 3$ at $x = 0$, the first can be substituted to obtain $B = 1-A$. Differentiating the general solution for y gives

$$dy/dx = -Ae^{-x} - 3Be^{-3x}$$

and substituting the second boundary condition then gives

$$3 = -A - 3B = 2A - 3$$

That is, $A = 3$ and $B = 2$.

branch: (jump) A departure front the normal sequential execution of instructions in a computer program. Control is thus transferred to another part of the program rather titan passing in strict sequence front one instruction to the next.

breadth: A horizontal distance, usually taken at right angles to a length.

British thermal unit: (Btu) A unit of energy equal to 1.055 06 ´ 10³ joules. It was formerly defined by the heat needed to raise the temperature of one pound of air-free water by one degree Fahrenheit at standard pressure.

buffer store: (buffer) A small area of the main store of a computer in which information can be stored temporarily before, during, and alter processing. A buffer can be used, for instance, between a peripheral device and the central processor, which operate at very different speeds.

bushel: A unit of capacity usually used for solid substances. In the UK it is equal to 8 UK gallons. In the USA it is equal to 64 US dry pints or 2150.42 cubic inches..

C

calibration: The marking of a scale on a measuring instrument. For example, a thermometer can be calibrated in degrees Celsius by marking the freezing point of water (0°C) and the boiling point of water (100°C).

calorie: Symbol: cal A unit of energy approximately equal to 4.2 joules. It was formerly defined as the energy needed to raise the temperature of one gram of water by degree Celsius. Because the specific thermal capacity of water changes with temperature, this definition is not precise. The mean or thermo-chemical calorie (call.) is defined as 4.184 joules. The international table calorie is defined as 4.186 8 joules.

cancellation: Removing a common factor in a numerator and denominator, or removing the same quantity from both sides of an algebraic equation. For example, xy/yz can be simplified by the cancellation of y, x/z. The equation $z + x = 2 + x$ is simplified to $z = 2$ by cancelling (subtracting) x front both sides.

candela: Symbol: cd The SI base unit of luminous intensity, defined as the intensity of b1ack-body radiation from a surface 1/600 000 square metre at the temperature of freezing platinum and at pressure of 101.325 pascals.

canonical form: (normal form) In matrix algebra the diagonal matrix derived by a series of transformation another square matrix of the same order.

capital: 1. The total sum of all the assets of a person or company, including cash, investments, household goods, land, buildings, machinery, and finished or unfinished goods.

2. A sum of money borrowed or lent on which interest is payable or received.

3. The total amount of money contributed by the shareholders when a company is formed, or the amount contributed to, a partnership by the partners.

carat: (metric) A unit of mass used for precious stones, It is equal in 200 milligrants.

card: A rectangular piece of stiff high quality paper on which information can be recorded. In the case of a punched card the informations recorded as a pattern of rectangular holes in the paper. Each punched card is a uniform size and is divided into a number of columns along its length, usually 80, The 80-column card measures 18.73 ´ 8.25 cm. Each of the 80 columns has 12 position at which a hole can be punched. A digit (0-9), letter, or some other character is represented by a particular combination of holes in a column. Thus a number of adjacent columns can be used to record an item of information.

Punched cards were the earliest means by which information could be fed into and obtained from a computer. Their use is however decreasing. The information is usually by means of a keypunch. This machine is manually operated from a keyboard similar to that of a typewriter which causes the necessary holes to be punched in each card column. The accuracy of the punching is checked by a machine known as a verifier. The punched information is then fed into the computer using a card reader. This device senses the presence or absence of holes in each column and convert this information into a series of electric pulses. The pulses are transmitted to the central processor of the computer. Although maybe 1000 cards can be rend in a minute, the card reader is considered a very slow input device. Information is output on punched cards by means of a card punch which automatically punches data into the cards. Compare paper tape, magnetic tape, disk.

cardinal numbers: Whole numbers that are used for counting, or for specifying the total number of items. That is, 1, 2, 3, ... Compare ordinal numbers.

cardioid: An epicycloid that has only one loop, formed by the path of a point on a circle rolling round the circumference of another that has the same radius.

Cartesian coordinates: A method of defining the position of a point by its distance front a fixed point (origin) in the direction of two or more straight fines, On a flat surface, two straight lines, called the x-axis and the y-axis, from the basis of a two-dimensional Cartesian coordinate system. The point at which they cross is the origin. An imaginary grid is formed by fines parallel to the axes and one unit length apart. The point (2,3), for example, is the point at which the fine parallel to the y-axis two units in the direction of the axis, crosses the line parallel to the x-axis three units in the direction of the y-axis. Usually the x-axis is horizontal and the y-axis is perpendicular to it. In three dimensions, a third, z-axis is added to define the height or depth of a Point. The coordinates of a point art then three numbers (x, y, z). A eight handed system is one for which if the thumb of the right hand points along the x-axis, the fingers of the hand fold in the direction in which the y-axis would have to rotate to point in the same direction as the z-axis. A left-handed system is the mirror image of this. In a rectangular system all three axes are mutually at right-angles.

catenary: The plane curve of a flexible uniform tint suspended from two points. For example, an empty washing line attached to two poles and hanging freely between them follows a catenary. The catenary is symmetrical about an axis perpendicular to the fine joining the two points of suspension. In Cartesian coordinates, the equation of a catenary that has its axis of symmetry lying along the y-axis, and that crosses the y axis at y = a, is

$$y = (a/2)(e^{x/a} + e^{-x/a})$$

catenoid: The curved surface formed by rotating a catenary about its axis of symmetry.

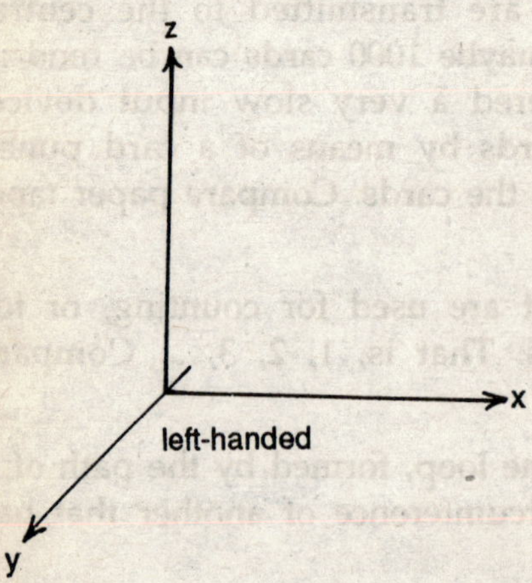

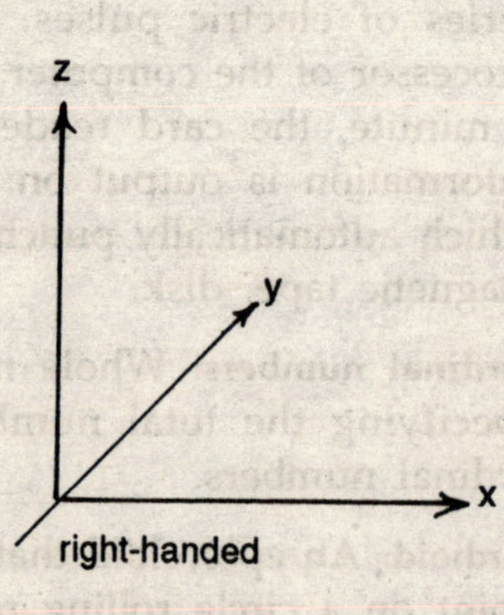

In three-dimensional Cartesian coordinates, right-handed system is the mirror image of a lefthanded system.

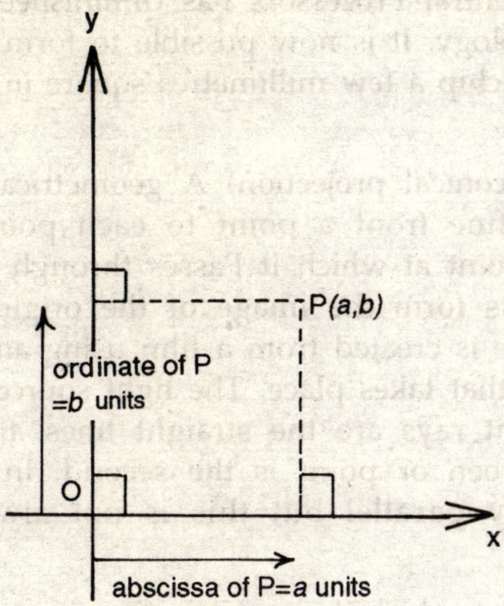

Two dimensional rectangular Cartesian coordinates showing a point P(a, b).

central conic: A conic with a centre of symmetry; e.g. an ellipse or hyperbola,

central enlargement: A central projection.

central force: A force that acts on any affected object along a line to an origin. For instance, the motion of electric forces between charged particles are central; frictional forces art not.

central processor: (central processing unit, CPU) A highly complex electronic device that is the nerve centre of a computer. It consists of the control unit and the arithmetic and logic unit (ALU). Also sometimes considered part of the central processor is the main store, or memory, where a program or a section of a program is stored in binary form. The control unit supervises all activities within the computer, interpreting the instructions that make up the program. Each instruction is automatically brought, in turn, from the main store and kept temporarily in a small store called a register. Electronic circuits analyse the instruction and determine the operation to be carried out and the exact location or locations in store of the data on which, the operation is to be performed, The operation is actually performed by the ALU, again using electronic circuitry and a set of registers. It could be an arithmetical calculation, such as the addition of two numbers, or a logical operation, such as selecting or comparing data. This process of fetching, analysing, and executing instructions is repeated in the required order until an instruction to stop is executed.

The size of central Processors has diminished considerably with advances in technology. It is now possible to form a central processor on a single Silicon chip a few millimetres square in area, or on a small number of chips.

central projection: (conical projection) A geometrical transformation in which a straight fine front a point to each point in the figure is continued to the point at which it Passes through the second (image) plane, These points form the image of the original figure. When a photographic image is created from a film using an enlarge this is the kind of Projection that takes place. The light source is at the centre of projection, the light rays are the straight lines, the film is the first plane, and the screen or point is the second. In this case the two planes are usually parallel but this is not always so in central projection.

centre: A point about which a geometric figure is symmetrical.

centre of buoyancy: (for an object in a fluid) The centre of mass of the displaced volume of fluid. For a floating object to be stable the centre of mass of the object must lie below the centre of buoyancy; when the object is in equilibrium, the two lie on a vertical fine.

centre of mass: A point in a body at which the whole mass of the body may be considered to act. Often the term centre of gravity is used. This is, strictly, not the same, except if the body is in a constant gravitational field. The centre of gravity is the point lit which the weight may be considered to act.

The centre of mass coincides with the centre of symmetry if the symmetrical body has a uniform density throughout. In other cases the principle of moments may be used to locate the point. For instance, two masses m_1 and m_2 a distance d apart have a centre of mass on the line between them, If this is a distance d_1 from m1 and *m2* from d_2 then

$$m_1d_1 = m2d2 \text{ or;}$$

$$m_1d_1 = m_2(d-d_1)$$

$$d_1 = m_2d/(m_1+m_2)$$

A more general relationship can be applied to a number of masses m1,m2,.....mi that are respectively distances r1,r2,........... ri from an origin. The distance r from the origin to the centre of mass is given by:

In the case of a body having a uniform density an integration must be used to obtain the position of the centre of mass, which coincides with the centroid.

centre of pressure: For a body or surface in a fluid, the point at which the resultant of pressure forces acts. If a surface lies horizontally in a fluid, the pressure at ail points will be the same. The resultant force will then act through the centroid of the surface. If the surface is not horizontal, the pressure on it will vary with depth. The resultant force will now act through a different point and the centre of pressure is net at the centroid.

centre of projection: The point at which ail the fines forming a central projection meet.

centrifugal force: A force supposed to act radially outwards on a body moving in a curve, In fact there is no real force acting; centrifugal force is said to be a ,fictitious, force, and the use is best avoided. The idea arises from the effect of inertia on an object moving in a curve. If a car is moving around a bend, for instance, it is forced in a curved path by friction between the wheels and the road. Without this friction the car would continue in a straight fine. The driver also moves in the curve, constrained by friction with the seat, restraint from a scat belt, or a 'push, from the door. To the driver it appears that there is a force radially outwards pushing his body out the centrifugal force. In fact this is not the case; if the driver tell out of the car he would move straight forward at a tangent to the curve. It is sometimes said that the centrifugal force is a reaction to the centripetal force this is net true.

centripetal force: A force that causes an object to move in a curved path rather than continuing in a straight line. The force is provided by, for instance: the tension of the string, for an object whirled on the end of a string; gravity, for an object in orbit round a planet; electric force, for an electron in the shell of an atom. The centripetal force for an object of mass m with constant speed v, and path radius r is mv^2/r, or mw2r, where w is angular speed. A body moving in a curved path has an acceleration because the direction of the velocity changes, even though the magnitude of the velocity may remain constant. This acceleration, which is directed towards the centre of the curve, is the centripetal acceleration. It is given by *v2 /r* or w^2r.

centroid: (mean centre) The point in a figure or solid at which the centre of mass would be if the figure or body were of uniform-density material. The centroid of a symmetrical figure is at the centre of symmetry; thus, the centroid of a circle is at its centre. The centroid of a triangle is the point at which the medians meet.

For nonsymmetrical figures or bodies integration is used to find the centroid. The centroid of a fine, figure, or solid is the point that has coordinates that are the mean values of the coordinates of ail the points in the fine, figure, etc. For a surface, the coordinates of the centroid are given by:

$$x = [\iint x dx dy]/A, \text{etc.}$$

the integration being over the surface, and A being the area. For a volume, a triple integral is used to obtain the coordinates of the centroid:

$$x = [\iiint x dx dy dz]V. \text{ etc}$$

c.g.s. system: A system of unit, that use the centimetre, the gram, and the second as the base mechanical units. Much early scientific work used ibis system, but it has now almost been abandoned.

chain: A unit of length equal to 22 yards, It is equivalent to 20.116 8m.

Chain rule: A rule for expressing the derivative of a function z f(x) in terms of another function of the same variable, u(x), where z is also a function of u. That is:

$$dz/dx = (dz/du)(du/dx)$$

This is often called the 'function of a function rule. For a function z = f(x1.x2,x3.........) of several variables, in each of the co-variables x1, *x2*, x3, ...is itself a function of a single variable t, the derivative, dz/dt called the total derivative is given by the chain rule for differentiation, which is:

$$dz/dt = (\partial x/\partial x_1)(dx1/dt) + (\partial z/\partial x_2)dx_2/dt + \ldots\ldots\ldots\ldots$$

channel: A Path along which information can travel in a computer system or communications system especially between the store and a magnetic tape or disk unit

character: One of a set of symbols that can be represented in a computer. It can be a letter, number, punctuation mark, or a special symbol. A character is stored or manipulated in the computer as a group of bits.

chi-square distribution: The distribution of the sum of the squares of random variables with standard normal distribution. For example, if x_1, x_2, ... are independent and all standard normal, then

$$X^2 = \Sigma x_i^2$$

has a chi-square distribution with n degrees of freedom written C_n^2. The mean and variance are n and 2n respectively. The values X_n^2 (a) for which $P(C^2$ £ X_n^2 (a)) = a or art tabulated for various values of n.

chi-square test: A measure, of how well a theoretical probability distribution fits a set of data. For i =1,2,3........m value xi occurs oi times in the data and the theory predicts that it will oocurs ei times. Provided ei ^{3}5 for all value of i then

$$X^2 = \Sigma(oi\ -ei)^2$$

has a chi-square distri-bution with n degrees of freedom.

chord: A straight line joining two points on a curve, for example, the line segment joining two Points on the circumference of a circle.

circle: The plane figure formed by a closed curve consisting of all the points that are at a fixed distance from a particular point in the plane; the point is the centre of the circle. The diameter of a circle is twice its radius; the circle is 2pr and its area is pr^2. In Cartesian coordinates, the equation of a circle centred at the origin is

$$x^2 + y^2 = r^2$$

The circle is the curve that encloses the largest Possible area within any given perimeter length.

circular cone: A cone that has i base.

circular cylinder: A cylinder in which the base is circular.

circular measure: The measurement of an angle in radians.

Circular motion: A form of periodic motion; that of an object moving in a circular path. For this to be possible, a positive central force must act. If the object has a uniform speed v and the radius of the circle is r, the angular velocity (w) is v^2/r. There is an acceleration towards the centre of the circle equal to v^2/r or w^2r.

circumcircle: (circumscribed circle) The circle that passes through all three vertices of a triangle or through the vertices of any other cyclic Polygon. The figure inside the circle is said to be inscribed. The point in the figure that is the centre of the circle is called the circumcentre. For a triangle with side lengths a, b, and c the radius r of the circumcircle is given by.

$$r = abc\ \sqrt{s(s\text{-}a)(s\text{-}b)(s\text{-}c)} \text{ where s is } (a + b + c)/2$$

circumference: The boundary, or length of the boundary, of a closed curve, usually a circle. The circumference of a is equal to 2pr, where r is the radius of the circle.

Circumscribed: Describing a geometric figure that is drawn around and enclosing another geometrical figure. For example, in a square, a circle can be drawn through the vertices. This is lied the circumscribed circle, and the square is called the inscribed square of the circle. Similarly, a regular polyhedron might have a circumscribed sphere, and a rectangular pyramid a circumscribed cone.

class: A grouping of data that is taken as one item in a frequency table or histogram.

classical mechanics: A system of mechanics that is based on Newton's laws of motion. Relativity effects and quantum theory are not taken into account in classical mechanics.

Clock pulse: One of a series of regular pulses that are produced by an electronic device called a clock and are used to synchronize operations in a computer. Every instruction in a computer program causes a number of operations to be done by the central processor of the computer. Each of these operations, performed by the control unit or arithmetic and logic unit, are triggered by one clock pulse and must be completed before the next clock pulse. The interval at which the pulses occur is usually a few microseconds.

clockwise: Rotating in the same sense as the hands of a clock. For example, the head of an ordinary screw is turned clockwise to drive the screw in. Looking at the other end the rotation appears to be anticlockwise.

closed: Describing a set for which a given operation gives results in the same set. For example the set of positive integers is closed with respect to addition and multiplica-tion. Adding or multiplying any two members gives another positive integer. The set is net closed with respect to division since dividing certain integers does not give a positive integer. The set of positive integers is also net closed with respect to subtraction.

Closed curve: (closed contour) A curve such as a circle or an ellipse, that forms a complete loop. It has no end points. A simple closed curve is a closed curve that does net cross itself.

closed interval: A set consisting of the numbers between two give a numbers (end points), including the end points. For example, ail the real numbers greater than or equal to 2 and less than or equal to 5 constitute a closed interval. The closed interval between two real numbers a and b is written [a, b]. On a number fine the end points are marked by a blacked-in circle.

closed set: A set in which the limits that define the set are included. The set of rational numbers greater than or equal to 0 and less than or equal to ten written and the set of points on and within a circle are examples of closed sets.

closed surface: A surface that has no boundary fines or curves, for example a sphere or an ellipsoid.

closed system: (isolated system) A set of one or more objects that may interact with each other, but do not interact with the world outside the system. This means that there is no net force from outside or energy transfer. Because of this the system's angular momentum, energy, mass, and linear momentum remain constant.

coding: The writing of instructions in a computer programming language. The person doing the coding starts with a written description or a diagram representing the task to be carried out by the computer. He then converts this into a precise and ordered sequence of instructions in the language that be has selected.

coefficient: A multiplying factor. For example, in the equation $2x^2 + 3x = 0$, where x is a variable, the coefficient of x is 2 and the coefficient of x, is 3. Sometimes the value of the coefficients is not known, although they are known to stay constant as x changes, for example, $ax^2 + bx = 0$. In this case a and b are constant coefficients.

cofactor: The determinant of the matrix obtained by removing the row and column containing the clement. The matrix formed by all the cofactors of the elements in a matrix is called the adjoint of the matrix.

coherent units: A system or subset of units in which the derived units are obtained by multiplying or dividing together base units, with no numerical factor involved.

collinear: Lying on the same straight fine. Any two points, for example, could be said to be collinear because there is a straight fine that passes through both. Similarly, two vectors are collinear if they are parallel and both act through the same point. column matrix

column vector: (column matrix) A number (m) of quantities arranged in a single column; i.e. an m x 1 matrix. For example, the vector that defines the displacement of the point (x, y, z) from the origin of a Cartesian coordinate system is usually written as a column vector.

combination: Any subset of a given set of objects regardless of the order in which they are selected. If r objects are selected from ri, and each object can only be chosen once, the number of different combinations is

$$n!/[\ r!(n-r)!]$$

written as nCr or C(n,r), For instance, if there are 15 students in a class and only 5 books, then each book has to be shared by 3 students. The number of ways in which this can happen— i.e the number of combinations of 3 from 15 is 15!/3!12!, or 455 The total number of possible subsets of any number is $nC_0 + nC_1 + \ldots\ldots\ldots nC_n$ which, from the binomial theorem, is 2^n. If each object can bc selected more than once the number of different combinations is $n+r-1C_r$..

commensurable: Able to be measured in the same way and in terms of the same units. For example, a 30 centimetre rule is commensurable with a 1 metre length of rope, because both can be measured in centimetres. Neither is commensurable with an area.

common denominator: A whole, number that is a common multiple of the denominators of two or more fractions. For example, 6 and 12 are both common denominators of 1/2 and 1/3. The lowest common denominator (LCD) is the smallest number that is a common multiple of the denominators of two or more fractions. For example, the LCD of 1/2, 1/3, and 1/4 is 12. Fractions are put in terms of the LCDs when they are to be added or subtracted:

$$1/2 + 1/3 + 1/4 = 6/12 + 4/12 + 3/12 = 13/12$$

common difference: The difference between successive terms in an arithmetic sequence or arithmetic series.

common factor: A whole number that divides exactly into two or more given numbers. For example, 7 is a common factor of 14, 49, and 84. Since 7 is the large number that divides into all three exactly, it is the highest common factor (HCF).

2. A number or variable by which several parts of an expression are multiplied. For example, in $4x^2 + 4y^2$, 4 is a common factor of x2 and y2 and from the distributive law for multiplication and addition

$$4x^2 + 4y^2 = 4(x^2 + y^2).$$

common multiple: A whole number that is a multiple of each of a group of numbers. For example, 100 is a common multiple of 5, 25, and 50. The lowest (or least) common multiple (LCM) is the smallest number that is a common multiple; in this case it is 50.

common ratio: The ratio of successive terms in a geometric sequence or geometric series.

common tangent: A single line that forms a tangent to two or more separate curves. The term is also used for the length of the fine joining the two tangential points,

commutative: Denoting an operation that is independent of the order of combination. A binary operation * is commutative if a*b = b*a for all values of a and b. In ordinary arithmetic, multiplication and addition are commutative operations. This is sometimes referred to as the commutative law of multiplication and the commutative law of addition. Subtraction and division art not commutative operations.

compasses: An instrument used for drawing circles. If consists of two rigid arms joined by a hinge. At one end is a sharp point, which is placed at the, centre of the circle, At the other end is a pencil or other marker, which traces out the circumference when the compasses are pivoted around the sharp point, In a beam compass, used for drawing large circles, the sharp point and the marker are attached to opposite ends of a horizontal beam.

complement: The set of all the elements that are not in a particular set. If the set A (1, 2 3), and the universal set, E is taken as containing a the natural numbers, then the complement of A, written A', is (4, 5, 6, ...).

complementary angles: A pair of angles that add together to make a right angle (90° or p/ 2 radians). Compare conjugate angles, supplementary angles.

completing the square: A way of solving a quadratic equation, by dividing both sides by the coefficient of the square term and adding a constant, in order to express the equation as a single squared term.

complex number: A number that has both a real part and an imaginary part. The imaginary part is a multiple of the square root of minus one (i). Some algebraic equations cannot be solved with real numbers. For example, $x^2 + 4x + 6 = 0$ has the solutions $x = -2 + \sqrt{(-2)}$ and $x = -2 - \sqrt{(-2)}$. If the number system is extended to include $i = \sqrt{-1}$ all algebraic equations can be solved. In this case the solutions are $x = 2 + i\sqrt{2}$ and $x = -2 - i\sqrt{2}$. The real part is -2 and the imaginary part is $+ i\sqrt{2}$ or $-i\sqrt{2}$.

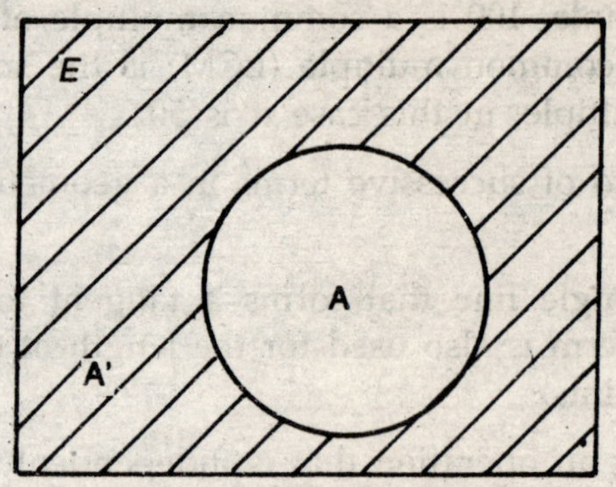

The shaded area in the venn diagram is the complement A' of the set A.

Complex numbers are sometimes represented on an Argand diagram, which is similar to a graph in Cartesian coordinates, but with the horizontal axis representing the real part of the number and the vertical axis the imaginary part. Any complex number can also bc written as a function of an angle θ just as Cartesian coordinates can be converted into polar coordinates. Thus r(cosθ + isinθ) is equivalent to x + iy, where x = rcosθ and y = rsinθ. Here, r is the modulus of the complex number and θ is the argument (or amplitude). This can also be written in the exponential form $r = e^{iq}$.

component vectors: The components of a given vector are two or more vectors with the same effect as the given vector. In other words the given vector is the resultant of the components. Any vector bas an infinite number of sets of components. Some sets are more use than others in a given case, especially pairs at 90°. The component of a given vector (V) in a given direction is the projection of the vector onto that direction; i.e. Vcosθ, where θ is the angle between the vector and the direction.

composite number: An integer that has more than one prime factor. For example, 4 (= 2 × 2), 6 (= 2 × 3). 10 (2 × 5) are composite numbers. The prime numbers and -1 are not composite.

compound interest: The interest earned on capital, when the interest in each period is added to the original capital as it is earned. Thus the capital, and therefore the interest on it, increases year by year. If P is the principal, R percent the interest rate pet annum, and n the number of interest periods, then the compound interest is

$$P(1 + R/100)^n$$

This formula is a geometric progression whose first term is P (when $n = 0$) and whose common ratio is $(1 + R/100)^n$.

computer: Any automatic device or machine that can perform calculations and other operations on data. The data must be received in an acceptable form and is processed according to instruction. The most versatile and most widely used computer is the digital computer, which is usually referred to simply as a computer. An automatically controlled calculating machine in which information, generally known as data, is represented by combinations of discrete electrical pulses denoted by the binary digits 0 and 1. Various operations, both arithmetical and logical, are performed on the data according to a set of instructions (a program). Instructions and data are fed into the main store or memory of the computer, where they are held until required. The instructions, coded like the data in binary form, are analysed and carried out by the central processor of the computer. The result of this processing is then delivered to the user. The technology used in digital computers is so highly advanced that they can operate at extremely high speeds and can store a huge amount of information. The thermionic valves used in early computers were replaced by transistors; transistors, resistors, etc., were subsequently packed into integrated circuits, which have become more and more complicated. As the electronic circuits used in the various devices in a computer system have decreased in size and increased in complexity, so the computers themselves have grown smaller, faster, and more powerful. The minicomputer and the even more compact microcomputer have been developed as somewhat simpler versions of the full-size mainframe computer. Computers now have an immense range of uses in science, technology, industry, commerce, education, and many other fields.

computer modelling: The development of a description or mathematical representation of a complicated process or system, using a computer. This model can then be used to study the behaviour or control of the process or system by varying the conditions in it, again with the aid of a computer,

concave: Curved inwards. For example, the inner surface of a hollow sphere is concave. Similarly in two dimensions, the inside edge of the circumference of a circle is concave. A concave polygon is a polygon that has one interior angles greater than 1801.

concentric: Denoting circles or spheres that have the same centre. For example, a hollowed out sphere consists of two concentric spherical surfaces. Compare eccentric.

condition: In logic, a proposition or statement, *P*, that is required to be true in order that another proposition *Q* be true. If *P* is a necessary condition then *Q* could not be true without *P*. If *P* is a sufficient condition, then whenever P is true *Q* is also true, but not vice versa. For example, for a quadrilateral to be a rectangle it must satisfy the necessary condition that two of its sides be parallel, but this is not a sufficient condition. A sufficient condition for a quadrilateral to be a rhombus is that all its sides have length of 5 centimetres, but this is not necessary condition. For a rectangle to be a square it is both a necessary and a sufficient condition that ail its sides are of equal length.

In formal terms, if *P* is a necessary condition for *Q*, then $Q \rightarrow P$, If *P* is a sufficient condition, then $P \rightarrow Q$, If P is a necessary and sufficient condition for *Q* then $P^{o}Q$.

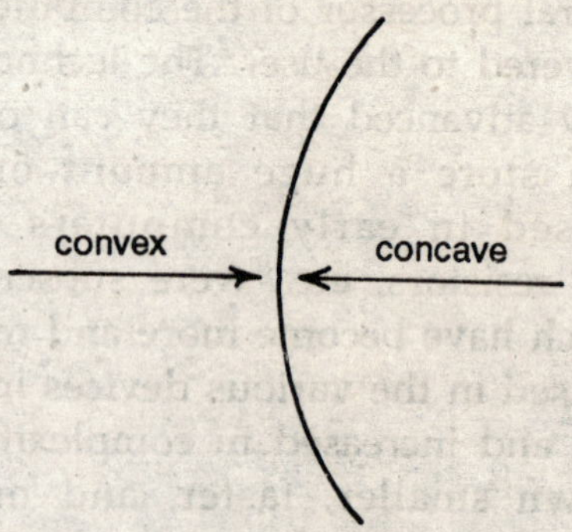

Concave and convex curvatures

cone: A solid defined by a closed plane curve (forming the base) and a point outside the plane (the vertex). A fine segment front the vertex to a point on the plane curve generates a curved lateral surface as the point moves around the plane curve. The fine is the generator of the cone and the plane curve is its directrix, Any litre segment front the vertex to the directrix is an element of the cone.

If the directrix is a circle the cone is a circular cone. If the base has a centre, a line from the vertex to this is an axis of the cone. If the axis is at right angles to the base the cone is a right cone; otherwise it is an oblique cone. The volume of a cone is one third of the base area multiplied by the altitude (the perpendicular distance front the vertex to the base). For a right circular cone

$$V = pr^2h/3$$

where r is the radius of the base and h the altitude. The area of the curved (lateral) surface of a right circular cone is prs. where s is the length of an element (the slant height). If an extended line is used to generate the curved surface (i.e. extending beyond the directrix and beyond the vertex), an extended surface is produced with two parts (nappes) each side of the vertex.

confidence interval: An interval that is thought, with a preselected degree of confidence, to contain the value of a parameter being estimated. For example, in a binomial experiment the a% confidence interval for the probability, of succes,, p lies between P a and P where

$$a = z\sqrt{[P(1-p)/N]}$$

N is the sample size, P the proportion of successes in the sample, and z is given by a table of area under the standard normal curve. P will lie in this interval a times out of every 100.

conformal mapping: A geometrical transformation that does not change the angles of intersection between two lines or curves. For example, Mercator's projection is a conformal mapping in which any angle between a fine on the spherical surface and a line of latitude or longitude will be the satire on the map.

congruent: Denoting two or more figures that arc identical in size and shape. Two congruent plane figures will fit into, the area occupied by each other; i.e. one could be brought into coincidence with the other by moving it without change of size. Two circles arc congruent if they have the same radius. The conditions for two triangles to, be congruent arc: (1) Two sides and thc included angle of one arc equal to two sides and the, included angle of the other. (2) Two angles and the included side of one arc equal to two angles and the included side of the other (3) Three sides of one are equal to three sides of the other. In solid geometry, two figures are congruent if they cm be brought into coincidence in space. Sometimes the term directly congruent is used to describe identical figures; indirectly congruent figures are ones that are mirror images of each other.

conic: A type of plane curve defined so that for all points on the curve the distance front a fixed point (the focus) has a constant ratio to the perpendicular distance from a fixed straight line (the directrix). The ratio is the eccentricity of the conic, e; i.e. the eccentricity is the distance from curve to focus divided by distance from curve to, directrix.

The type of conic depends on the value of e: when e is less than 1 it is an ellipse; when e equals 1 it is a parabola; when e is greater than 1 it is a hyperbola.

The original definition of conics was as plane sections of a cortical surface—hence the name conic section. In a cortical surface having an apex angle of 2q the cross-section on a plane that makes an angle q with the axis of the cone, is a parabola. A cross-section in a plane that makes an angle less than q with the axis is an ellipse. A cross-section in a plane making less than q with the axis is a hyperbola and because this plane cuts both halves (nappes) of the cone, the hyperbola has two arms.

There are various ways of writing the equation of a conic. In Cartesian coordinates:

$$(1-e^2)x^2 + 2e^2qx + y^2 = e^2q$$

where the focus is at the origin and the directrix is the line $x = q$. The general equation of a conic is $ax^2 + bxy + cy^2 + dx + ey + f = 0$ where a, b, c,d, e, and f are constants. This includes degenerate cases such as a point, a straight line, and a pair of intersecting straight lines. A point, for example, is a section through the vertex of the cortical surface. A pair of intersecting straight lines is a section down the axis of the surface. The tangent to the general conic at the point (x_1, y_1) is: $ax_1x + b(xy_1 + x_1y) + cy_1y + d(x + x_1) + e(y + y_1) + f = 0$

conjugate angles: A pair of angles that add together to make a complete revolution (360° or 2p radian)

conjugate complex numbers: Two complex numbers of the form $x + iy$ and $x - iy$, which when multiplied together have a real product $x^2 + y^2$. If $z = x + iy$, the complex conjugate of z is $z¢ = x - iy$.

conjunction: In logic, the relationship and between two or more propositions or statements. The conjunction of P and Q is true when P is true and Q is true, and false otherwise. The truth table definition for conjunction is shown in the illustration.

connectivity: The number of cuts needed to break a shape in two parts. For example, a rectangle, a circle, and a sphere, all have a connectivity of one. A flat disc with a hole in it or a torus has a connectivity of two.

consequent: In logic, the second part of a conditional statement; a proposition or statement that is said to follow front or be implied by, another. For example, in the statement 'if Jill is happy then Jack is happy 'Jack is happy' is the consequent.

conservation law: A law stating that the total value of some physical quantity is conserved throughout any changes in closed system. The conservation applying in mechanics arc the laws of constant mass, constant energy, constant linear momentum, and constant angular momentum.

conservation of mass and energy: The law that the total energy (rest mass energy + kinetic energy + potential energy) of a closed system is constant. In most chemical and physical interactions the mass change is undetectably small, so that the measurable rest-mass energy does not change. The law then becomes the classical law of conservation of energy. In practice, the inclusion of mass in the calculation is necessary only in the case of nuclear changes or systems involving very high velocities.

conservative field: A field such that the work done in moving an object between two points in the field is independent of the path taken.

conservative force: A force such that, if it moves an object between two points, the energy transfer does not depend on the path between the points. It must then lie true that if a conservative force moves an object in a closed path, the energy transfer is zero. Gravitation is an example of a conservative force; friction is a nonconservative force.

consistent: Giving rise to no contradictions. Arithmetic, for example, is a consistent logical system because none of its axiom nor any of the theorems that derived from these by the rules, are contradictory,

consistent equation: A set of equations that can be satisfied by at least one set of values for the variables. For example, the equations $x + y = 2$ and $x + 4y = 6$ are satisfied by $x = 2/3$ and $y = 4/3$ and they are therefore consistent. The equations $x + y = 4$ and $x + y = 9$ are inconsistent.

constant: A quantity that does not change its value in a general relationship between variables. For example, in the equation $y= 2x + 3$, where x and y are variables, the numbers 2 and 3 constants. In ibis case they are absolute constants because their values never change. Sometimes a constant car, take any one of a number of values in different applications of a general formula. In the general quadratic equation

$$ax^2+ bx + c = 0$$

a, b, and c are arbitrary constants because no values are specified for them.~ An indefinite integral includes an arbitrary constant because for a function f(x) the integral with respect to x has the form

$$\int f(x)dx + c$$

where the value of the constant c depends on the limits chosen.

constant angular momentum law of: The principle that the total angular momentum of a system cannot change unless a net outside torque acts on the system.

constant energy, law of: (law of conservation of energy) The principle that the total energy of a system cannot change unless energy is taken from or given to the outside.

constant linear momentum, law of: (law of conservation of (linear) momentum) The principle that the total linear momentum of a system cannot change unless a net outside force acts.

constant mass law of: The principle that the total mass of a system cannot change unless mass is taken from or given in the outside.

continued product: Symbol: P The product of a number of related terms. For example, 2 x 4 x 6 x 8...... is a continued product, written:

$$P^k an$$

This means the product of k terms, with the nth term, a = 2n.

$$P_1^{¥} an$$

has an infinite nurnber of term is.

continuous function: A function that has no sudden changes in value as the variable increases or decreases smoothly. More precisely, a function f(x) is continuons at a point x a if the limit of f(x) as x approaches a is f(a). When a function dots not satisfy this condition at a point, it is said to be discontinuous, or to have a discontinuity, at that point. For example, tanq has discontinuities at q = p/2, 3p/2,5p/2, ... A function is continuous in an interval of x if there are no points of discontinuity in that interval.

contour line: A line on a map joining points of equal height. Contour lines are usually drawn for equal intervals of height, so that the steeper a slope, the closer together the contour fines.

contradiction: In logic, a proposition, statement, or sentence that both asserts something and denies it. It is a form of words or symbols that cannot possibly be true; for example, 'if I can read the book then I cannot read the book' and 'he is coming and he is not coming.

contrapositive: In logic, a statement in which the antecedent and consequent of a conditional am reversed and negated. The contrapositive of A→B is~B→A and the two statements are logically equivalent

convergent sequence: A sequence in which the difference between each term and the one following it becomes smaller throughout the sequence; i.e. the difference between the nth term and the (n + 1)th term decreases as n increases. For example, (1, 112, 1/4, 1/8, ...) is a convergent sequence, but (1, 2,4, 9, ...) is not. A convergent sequence has a limit a value towards which the nth term tends as n becomes infinitely large.

convergent series: A series in which the sum of all the term is after the nth term becomes smaller as n increases. For example.

$$Sn = 1 + 1/2 + 1/4 + 1/8 + ... + 1/2^n ...$$

is a convergent series. For every convergent series there is a sum to infinity, which is the limit of the sum of n term is as n tends te infinity. In this case the sum to, infinity is 2.

converse: A logical implication taken in the reverse order. For example, the converse of

if I am under 16, then I go to school is

if I go to, school, then 1 am under 16.

The converse of an implication is not always true if the implication itself is true. There are a number of theorems in mathematics for which both the statement and the converse are true. For example, the theorem: if two chords of a circle are equidistant front the centre then they are equal has a true converse: if two chords of a circle are equal, then they are equidistant from the centre of the circle.

conversion factor: The ratio of a measurement in one set of units to the equivalent numerical value in other units. For example, the conversion factor front inches to centimetres is 2.54 because 1 inch = 2.54 centimetres.

conversion graph: A graph showing a relationship between two variable quantities. If one quantity is known, the corresponding value of the other can be read directly front the graph. For example air pressure depends on height above sea level, A standard curve of altitude against air pressure may be plotted on a graph. An air-pressure measurement can be converted to an indication of height by reading the appropriate value from the graph,

convex: Curved outwards. For example, the outer surface of a sphere is convex. Similarly, in two dimensions, the outside of a circle is its convex side. A convex polygon is one in which no interior angle is greater than 180'.

coordinates: Numbers that define the position of a point, or set of points. A fixed point, called the origin, and fixed fines, called axes are used as a reference, For example, a horizontal line and a vertical line drawn on a page might be defined as the x-axis and the y-axis respectively, and the point at which they cross as the origin (O). Any point on the page can then be given, number its distance from O along the x-axis from left to right and its distance upwards from O in the direction of the y-axis. These two numbers would be the x and y coordinates of the point. This type of coordinate systems known as a rectangular Cartesian coordinate system. It can have two axes, as on a flat surface, such as a map, or three axes, when depth or height also have to, be specified. Another type of coordinate system (polar coordinates), expresses the position of a point as radial distance from the origin (the pole), with this direction expressed as an angle, or angles (positive when anticlockwise), between the radius and a fixed axis (the polar axis). Sec also, Cartesian coordinates, polar coordinates.

coplanar: Lying in the same plane, Any set of three points, for example, could be said to be coplanar because there is a plane in which they ail lie. Similarly, two vectors are coplanar if there is a plane that contains both.

coplanar forces: Forces in a single plane. If only two forces act through a point, they must be coplanar. So too are two parallel forces. However, nonparallel forces that do not act through a point may not be coplanar. If a set of coplanar forces act on a body their algebraic sum must be zero. In addition there must be no couple on the body.

coriolis force: A fictitious' force used to describe the motion of an object in a rotating system. For instance, air moving front north to south over the surface of the Earth would, to an observer outside the Earth, be moving in a straight line. To an observer on the Earth the path would appear to be curved, as the Earth rotates. Such systems can be described by introducing a tangential Coriolis 'force'. The idea is used in meteorology to explain wind directions.

correlation coefficient: A measure of the degree to which there is a linear relationship between two variables x and y. Given a set of data (xi,> 1), (x2,y2), ... (x_n, y_n), the correlation coefficient r_{xy} is equal to the covariance of x and y divided by (standard deviation of x x standard deviation of y). It is S_{xy}/S_xS_y

If the modulus of r is close to one, x and y are highly correlated and a graph of y against x shows, that the points (x1, y1), (x2, y2)....... .. (xn, yn) lie almost on a straight line.

If r is positive, x and y are said to bc positively correlated; i.e. as x increases y increases.

If r is negative they are inversely correlated; as x increases y decreases.

cosecant: (cosec, csc) A trigonometric function of an angle equal to the reciprocal of of its sine; i.e. coseca = 1 /sina.

cosech A hyperbolic cosecant.

cosh A: hyperbolic cosine.

cosine: (cos) A trigonometric function of an angle.

The cosine of an angle a (cosa) in a right-angled triangle is the ratio of the side adjacent to it, to the hypotenuse.

This definition applies only to angles between 0° and 90°.

More generally, in rectangular Cartesian coordina-tes, the x-coordinate of any point on the circumference of a circle of radius r centred at the origin is rcosa where a is the angle between the x-axis and the radius to that point.

In other words, the cosine function is the horizontal component of a point on a circle.

Cosa varies periodically in the same way as sina, i.e. but 90° ahead that is cosa is 1, when a is 0° falls to 0 when a = 90° and then to -1 when a =180° and returning to 0 when a = 270° and then to +1 again at a 360 °.

This cycle is repeated every complete revolution.

The cosine function has the following properties:

$$\cos a = \cos(a + 360°) = \sin(a + 90°)$$

$$\cos a = \cos(-a)$$

$$\cos(90° + a) + -\cos a$$

The cosine function can also, be defined as an infinite series. In the range front + 1 to -1:

$$\cos x = 1 - x^2/2! + x^4/4! - x^6/6! + \ldots\ldots\ldots$$

cosine rule: In any triangle, if a, b, and c are the side lengths and g is the angle opposite the side of length c, then

$$c^2 = a^2 + b^2 - 2ab\cos g$$

cotangent: A trigonometric function of an angle equal to the reciprocal of its tangent; i.e. cota = 1 /tana

coth: A hyperbolic cotangent.

coulomb: Symbol: C The SI unit of electric charge, equal to the charge transported by an electric current of one ampere flowing for one second. 1 C = 1A s.

couple: A pair of equal parallel forces in opposite directions and not acting through a single point. Their linear resultant is zero, but there is a net turning effect (moment). The net turning effect T (the torque) is given by:

$$T = Fd_1 + Fd_2$$

F being the magnitude of each force and dl and dl the distances from any point to the lines of action of each force. This is equivalent to:

$$T\ Fd$$

where d is the distance between the forces.

covariance: A statistic that measures the association between two variables. If, for x and y there are n pairs of values (xi, yi) then the covariance is defined as

[1/(n-1)]S(xi-x¢)(yi-y¢)]

where x¢and y' are the mean values.

critical path: The sequence of operations that should be followed in order' to complete a complicated process, task etc., in the minimum time. It is usually~ determined by using a computer.

cross section: (section) A plane cutting through a solid figure or the plane figure produced by such a cut. For example, the cross section through the middle of a sphere is a circle. A vertical cross section through an upright cone is a hyperbola.

cube: 1. The third power of a number or variable. The cube of x is x x ´ x = x^3 .

2. In geometry, a solid figure that has six square faces. The volume of a cube is l^3 where 1 is the length of a side.

cube root: An expression that has a third power equal to a given number. 'Me cube root of 27 is 3, since $3^3 = 27$.

cubic equation: A polynomial equation in which the highest power of the unknown variable is three. The general form of a cubic equation in a variable x is $ax^3 + bx^2 + cx + d = 0$

where a, b, c, and d are constants. It is also sometimes written in the reduced form

$$x^3 + bx^2/a + cx/a + d/a = 0$$

In general, there are three values of x that satisfy a cubic equation.

cuboid: A box-shaped solid figure bounded by %ix rectangular faces. The opposite faces are congruent and parallel. At each of the eight vertices, three faces meet at right-angles to each other. The volume of a cuboid is its length, l, times its breadth, b, times its height, h. The surface area is the sum of the areas of the faces, that is

$$2(l \times b) + 2(b \times h) + 2(l \times h)$$

In the special case in which l= b = h, all the faces are square and the cuboid is a cube of volume l^3 and surface area l^2

cumulative frequency: The total frequency of all values up to and including the upper boundary of the class interval under consideration.

curl: (SymbolÑx) A vector operator on a vector function that, for a three-dimensional function, is equal to the sum of the vector product (cross product) of the unit vectors and partial derivatives in each of the component directions. That is:

$$\text{curl } F = \nabla \times F = i \times \partial F/\partial x + j \times \partial F/\partial y + k \times \partial F/\partial z$$

where i, j, and k are the unit vectors in the x. Y, and z directions respectively. In Physics, the curl of a vector arises in the relationship between electric current and magnetic flux, and in the relationship between the velocity and angular momentum of a moving fluid.

curve: A set of points that form or can be joined by a continuous fine on a graph or other surface curve. A set of Points forming a continuous line. For example, in a graph plotted in Cartesian coordinates, the curve of the equation $y = x^2$ is a parabola. A curved surface may similarly represent a function of two variables in three-dimensional coordinates.

cusp: A sharp point formed by a discontinuity in a curve. For example two semicircles placed side by side and touching form a cusp at which they touch.

cybernetics: The branch of science concerned with control systems, especially with regard to the comparisons between those of machines and those of man and other animals. In a series of operations, information gained at one stage can be used to modify later performances of that operation. This is known as feedback and enables a control system to check and possibly adjust its actions when required.

cycle: A series of events that is regularly repeated. A cycle is a complete single set of changes, starting from one point and returning to the same point in the same way.

cyclic group: A group in which each element can be expressed as a power of any other element.

cyclic polygon: A polygon for which there is a circle on which all the vertices lie. All triangles are cyclic, All regular polygons are cyclic. All squares and rectangles are cyclic quadri-laterals. However, not all quadrilaterals are cyclic. Convex quadrilaterals are cyclic if the opposite angles are supple-mentary. For a cyclic quadri-lateral with sides of length a, b, c, and d (in order) the expression (ac + bd) is equal to the product of the diagonals. This is known as Ptolemy's theorem.

cycloid: The curve traced a point on a circle rolling along a straight line, for example, a point on the rim of a wheel roiling along the ground. For a circle of radius r rolling along a horizontal axis, the cycloid produced is a series 01 continuous arcs that rises front the axis to a height 2r and fall to touch the axis again at a cusp point, where the next arc begins. The horizontal distance between successive cusps is 2pr, the circumference of the circle. The length of the cycloid between adjacent cusps is 8r. If & is the angle formed by the radius to a point P (x, y) on the cycloid and the radius to the point of contact with the x-axis, the parametric equations of the cycloid are:

$$x = r(\theta - \sin\theta)$$

$$y = r(\theta - \cos\theta)$$

cylinder: A solid defined by a closed plane curve (forming a base) with an identical curve parallel to it.

Any line segment from a point on one curve to, a corresponding point on the other curve is art element of the cylinder.

If one of these elements moves parallel to itself round the base it sweeps out a curved lateral surface. The line is a generator of the cylinder and the plane closed curve forming that base is called the directrix. If the bases are circles the cylinder is a circular cylinder.

If the bases have centre the line joining them is an axis of the cylinder. A right cylinder is one with its axis at right angles to the base; otherwise it is an oblique cylinder.

The volume of a cylinder is Ah, where A is the base area and h the altitude (the perpendicular distance between the bases). For a right circular cylinder, the curved lateral surface area is 2prh, where r is the radius.

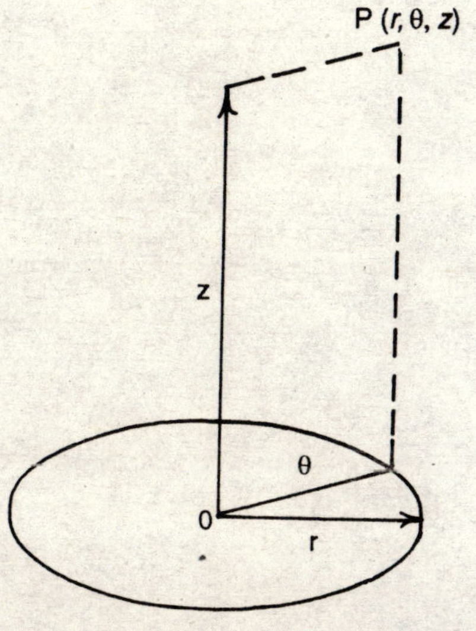

A point p(r, θ, z) in cylindrical polar

cylindrical polar coordinates: A method of defining the position Of a Point in space by its horizontal radius r from a fixed vertical axis, the angular direction 0 of the radius front an axis, and the height z above a fixed horizontal reference plane, Starting at the origin O of the coordinate system, the point p(r,θ,z) is reached by moving out along a fixed horizontal axis to a distance r, following the circumference of the horizontal circle radius r centred at O through ail angle q, and the" moving vertically upwards by a distance z. For a point P(r, θ, z), the corresponding rectangular Cartesian coordinates (x, ψ, z) are:

$$x = r\cos\theta$$

$$y = r\sin\theta$$

$$z = z$$

Compare spherical polar coordinates see also cartesian coordinates, coor dinates, polar coordinates.

D

damped oscillation: An oscillation with an amplitude that progressively decreases with time.

damping: The reduction in amplitude of a vibration with time by some form of resistance. A swinging pendulum will at last come to rest; a plucked string will net vibrate for long in both cases internal and/or external resistive forces Progressively reduce the amplitude and bring the system to equilibrium, In many cases the damping force(s) will be proportional to the object's speed. In any event, energy must be transferred from the vibrating system, to overcome the resistance. Where damping is an asset, the optimum situation occurs when the motion comes to zero in the shortest time possible, without vibration. 'This is critical damping. If the resistive force is such that the time taken is longer than this, over-damping occurs. Conversely, under-damping involves a longer time with vibrations of decreasing amplitude.

data: (now often used as a singular noun) The facts that refer to or describe an object, idea, condition, situation, etc. In computing, data can be regarded as the facts on which a program operates as opposed to the instructions in the Program. It can only be accepted and Processed by the computer in binary form. Data is sometimes considered to be numerical information only.

data bank: A large collection of organized computer data from which particular pieces of information can be ready extracted.

database: A large collection of organized data providing a common pool of information for users, say, in the various sections of a large organization. Information can be added, deleted and updated as required. The management of a database is very complicated and costly so that computer programs have been developed for this purpose. These programs allow the information to be extracted in many different ways. For example, a request could be put in for an alphabetical list of men over a certain age and living in ?, specified area, in which their employment and income should be given. Alternatively the request could be for an alphabetical fist of men over a certain age and income level in which their address and form of employment should be given.

data processing: The sequence of operations performed on data in order to extract information or to achieve same form of order. The term usually means the processing of data by computers but can also include its observation and collection.

debug: To detect, locate, and correct errors or faults (bugs) that occur in computer programs or in pieces of computer equipment. Since programs and equipment are often highly complicated, debugging can be a tedious and lengthy job. Programming errors may result from the incorrect coding of an instruction or from using instructions that will net give the *required solution* to a problem. Syntax errors can usually be detected and located by the compiler logic errors can be more difficult to find.

decagon: A plane figure with ten straight sides. A regular decagon has ten equal sides and ten equal angles of 36°

decahedron: A polyhedron that has ten faces.

deci: Symbol: d A prefix denoting 10'. For example, 1 decimetre (dm) = 10^{-1} metre (m).

decibel: Symbol: dB A unit of power level, usually of a sound wave or electrical signal, measured on *P* logarithmic scale. The threshold of hearing is taken as 0 dB in sound measurement.

Ten times this power level is 10 dB. The fundamental unit is the *bel*, but the decibel is almost exclusively used. A power P has a power level in decibels given by:

$$10 \log_{10}(P/ P0)$$

where P_0 is the reference power.

decimal: Denoting or based on the number ten. The numbers in common use for counting form a decimal number system.

A decimal rational number written a, units tenths, hund-redths, thousandths and so on.

For example, 1/4 is 0.25 in decimal notation. This type of decimal fraction is a finite decimal because the third and subsequent digits alter the decimal point art 0. Some rational numbers, such as5/27 (0.185 185 185.....) cannot be written as an exact decimal, but result in a number of digits that repeat indefinitely. These are called repeating decimals. All rational numbers can be written as either finite decimals or repeating decimals.

A decimal that is net finite and does not repeat is an irrational number and can be quoted to any number of decimal places, but never exactly.

deduction: A series of logical steps in which a conclusion is reached directly front a set of initial statements.

A deduction is valid if a sentence or statement that asserts the premisses and denies the conclusion is a contra-diction.

definite integral: (Riemann integral) The result of inte-grating any function of a single variable, f(x), between two specified values of x: x_1 and x2. The definite integral of f(x) is written as

$$\int_{x1}^{x2} f(x)dx$$

If the general expression for the integral of f(x) is another function of x, g(x), the definite integral is given by:

$$g(x_1) - g(x_2)$$

definition: In a measurement, the accuracy with which the instrument reading reflects the true value of the quantity being measured.

deformation: A geometrical transformation that stretches, shrinks, or twists a shape but does net break up any of its lines or surfaces.

It is often called, more precisely, a continuous defor-mation.

degree: 1. Symbol: ° A unit of plane angle equal to one ninetieth of a right angle.

2. A unit of temperature.

3. The exponent of a variable.

For instance $3x^3$ has a degree of 3. If there are several term is the sum of the exponents is used $5xy^2z^3$ has a degree of 6.

4. The highest power of a variable in a polynomial. For example in $x^3+ 2x + 1$, the degree is 3.

5. The highest power of an equation. For example, the degree of the equation $x^4 + 2x = 0$ is 4.

6. The highest power to which the highest order derivative is raised in a differential equation.

For example, the degree of $(d^2y/dx^2)^3 + dy/dx = 0$ is three. The degree of $d3y/dx3 + 2y(d^2y/dx^2)^2 = 0$ is one.

De l'Hospitals rule: The rule stating that the limit of the ratio of two functions of the same variable (x) as x approaches a value a, is equal to the limit of the ratios of their derivatives with respect to x.

That is, the limit of $f(x)/g(x)$ as $x \to a$ is the limit of $f'(x)/g'(x)$ as $x \to a$.

De l'Hopital's rule can be used to find the limits of $f(x)/g(x)$ at points at which both $f(x)$ and $g(x)$ are zero and the ratio is therefore indeterminate.

Any function that gives rise to an indeterminate form, and that can be expressed as a ratio of two functions, can be dealt with in this way.

For example, in

$$F(x) = (x^2-3)/(x\ 3)$$

writing

$$f(x)\ (x^2-3)$$

and

$$g(x) = (x - 3)$$

$F(x) = f(x)/g(x)$ The limit of $F(x)$ as $x \to 3$ is indeterminate . It can be obtained by using the limit of

$$f'(x)/g'(x) = 2x$$

as $x \to 3$. Thus the limit is 6.

De Moivre's theorerm: A formula for calculating a power number. If the number is in the polar form

$$z = r(\cos\theta + i\sin\theta)$$

$$\text{then } z^n = r^n(\cos n\theta + i\sin n\theta)$$

denominator: The bottom, part of a fraction. For example, in the fraction 3/4, 4 is the denominator and 3 is the numerator. The denominator is the divisor.

denumerable set: A set in which the elements can be counted. For example, the set of prime numbers, although infinite, can be counted, as can the set of positive even integers. These are known as denumerably infinite sets. The set of rational numbers, on the other hand, is not denumerable because be*t*ween any two elements there can always be found a third.

deposit: A sum of money paid by a buyer, either to reserve goods or property that he wishes to buy at a later date or as the first of a series of instalments in a hire-purchase agreement. If the buyer fails to complete the purchase the deposit is normally forfeited.

depth: The distance downwards from a reference level or backward from a reference plane. For example, the distance below a water surface and the distance between a wall surface and the back of an alcove in the wall, are both called depths derivative The result of differentiation.

derived unit: A unit defined in terms of base units, and not directly from a standard value of the quantity measures. For example, the newton is a unit of force defined as a kilogram metre second^{-2} (kg m s^{-2}).

determinant: A function of a square matrix derived by multiplying and adding the elements together to obtain a single number. For example, in a 2 x 2 matrix the determinant is $a_1b_2 = a_2b_1$. This is written as a square array in vertical lines, symbol D_2 and is called a second order determinant. Determinants occur in simultaneous equations. The solution of

$$a_1x + b_1y = c_1$$

A third order determinant has three rows and columns, and arises in a similar way in sets of three simultaneous equations in three variables.

The determinant of a transpose of a matrix, ½A ½ is equal to the determinant of the matrix, ½A½. If the position

developable surface: A surface that can be rolled out flat onto a plane. The lateral surface of a cone.

diagonal: Joining opposite corners. A diagonal of a square, for example, cuts it into two congruent right angled triangles. In a solid figure, usually a polyhedron, a diagonal plane is one that passes through two edges that are not adjacent.

diagonal matrix: A square matrix in which ail the elements are zero except those on the leading diagonal, that is, the first element in the first row, the second element in the second row, and so on. Diagonal matrices, unlike most others, are commutative in matrix multiplication.

diameter: The distance across a plane figure or a solid at its widest point. The diameter of a circle or a sphere is twice the radius.

diametral: Denoting a line or plane that forms a diameter of a figure. For example a cross section through the centre of a sphere is a diametral plane.

difference The result of subtracting one quantity or expression front another.

differential: An infinitesimal change in a function of one of more variables, resulting from a small change in the variables. For example, if f(x) is a function of x, and f changes by as a result of a change Δx in x, the differential, df, is defined as the limit of Δf as Δx beccomes infinitely small, That is $df = f'(x)dx$, where $f'(x)$ is the derivative of f with respect to x. This is a total differential, because it takes into account changes in all of the variables. For a function of two variables, f(x,y) the rate of change of f with respect to x is the partial derivative $\partial f/\partial x$. The change in f resulting from Changing x by dx and keeping y constant is the partial differential, $(\partial f/\partial y)dy$. For any function, the total differential is the some of all the partial differentials. For f(x,y); $df = (\partial f/\partial x).dx + (\partial f/\partial y).dy$.

differential equation: An equation that contains derivatives. An example of a simple differential equation is: $dy/dx + 4x + 6 = 0$

To solve such equations it is necessary to use integration. The, equation above can be rearranged to give:

$$dy, = -(4x + 6)dx$$

integrating both sides

$$\int dy = \int -(4x + 6)dx$$

which gives:

$$y = -2x^2 + 6x + C$$

where C is a constant of integration. The value of C can be found if particular values of x and y are known: for example, if y = 1 when x = 0 then C 1, and the full solution is

$$y = -2x^2 - 6x + 1$$

Note that the solution to a differential equation is itself an equation. Differentiating the solution gives the original equation. Equations like that above, which contain only first derivatives (dy/, dx) are said to be first order, if they contain second derivatives they are second order in general, the order of a differential equation is the highest derivative in the equation. The degree of a differential equation power of the highest order derivative. The differential equation in the example given is a first order and first degree equation. It is an example of a type of equation solvable by separating the variables onto both sides of the equation, so that each can be integrated. Another type of first-order first-degree equation is one of the form:

$$dy/dx = f(y/x)$$

Such equations are known as homogeneous differential equations. An exam pie is the equation:

$$dy/dx \ (x^2+y^{2'})/x^2$$

To solve homogeneous equations a substitution is made -y = mx, where m is a function of x. Then:

$$dy/dx = m + xdm/dx$$

and

$$(x^2 + y^2)/x^2 = (x\ ^2+ m^2x^2)/x^2$$

So the equation becomes:

$$m + xdm/dx = (x\ ^2+ m^2x^2)/x^2$$

or.

$$xdm/dx = 1 + m^2\text{-}m$$

The equation can now be solved by separating the variables.

An equation of the form:

$$dy/dx + P(x)y = Q(x)$$

where P(x) and Q(x) are functions of x (but not y), is a linear differential equation. Equations of this type can be put in a solvable form by multiplying both sides by the expression:

$$\exp(\int P(X)dx)$$

This is known as an integrating factor. For example, the differential equation

$$dy/dx + y/x = x^2$$

is a linear first-order differential equation. The function P(x) is 1/x, so the integrating factor is:

$$\exp(\int dx/x)$$

which is exp(logx). i.e. x Multiplying both sides of the equation by x gives:

$$xdy/dx + y = x^3$$

The left-hand side of the equation is equal to d(xy)/dx, so the equation becomes

$$d(xy)/dx = x^3$$

Integrating both sides gives:

$$xy = x^4/4 + C$$

where C is a constant.

differentiation: A process finding rate at which one variable quantity changes with respect to another. For example, a car might travel along a road from position x_1 to position x_2 in a time interval t_1 to t_2 Its average velocity is $(x_2 - x_1)/(t_2 - t_1)$ which can be written $\Delta x/\Delta t$ where Δx represents the change in x in the time Dt. However, the car might accelerate or decelerate in this interval and it may be necessary to know the velocity at a particular instant say t_1. In this case the time interval Δt is made infinitely small, i.e. t_2 can be as close as necessary to t_1. The limit of $\Delta x/\Delta t$ as Δt approaches zero is the instantaneous velocity at t_1 The result of differentiation, or derivative of a function y = f(x) is written d/ldx or f′(x). On a graph f(x), dy/dx at any point is the slope of the tagent to the curve y = f(x) at that point.

digit: A symbol that forms part of a number. *For e*xample, in the number 3121 there four digits. The ordinary number system has ten digits (0-9), where as binary (base two) system needs only two, 0 and 1.

digital: Using numerical digits. For example, a digital watch shows the time in numbers of hours and minutes and net as the position of hands on a dial. In general, digital devices work by some kind of counting process, either mechanical or electronic. The abacus is a, very simple example. Early calculating machines counted with mechanical relays. Modern calculators use electronic switching circuits.

dihedral: Formed by two initersecting planes, Two planes intersect along, a straight line (edge). The dibedral angle (or dihedron) between the planes is the angle between two fines (one in each plane) drawn perpendicular to the edge from a point on the edge. The dihedral angle of a polyhedron is the angle between two &faces.

dilatation: A geometrical mapping or projection in which a figure is stretched', not necessarily by the same amount in each direction. A square, for example, may be mapped into a rectangle by dilatation, or a cube into a cuboid.

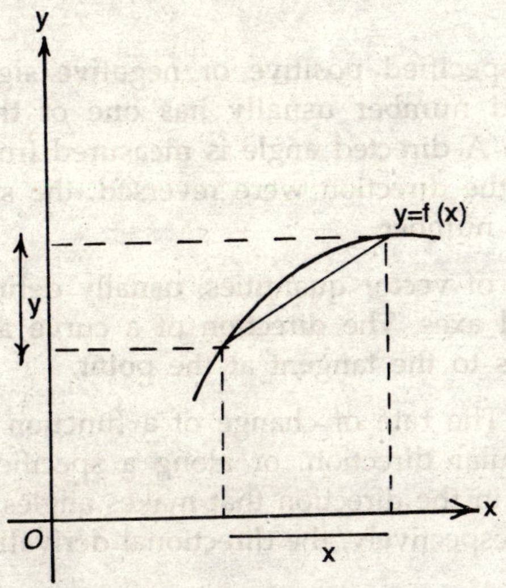

Differentiation of a function y = f (x). the derivative dy/dx is the limit of Δy/Δx as Δx and Δy become infinitely small.

dimension: The number of coordinates needed to represent the points on a line, shape, or solid. A plane figure is said to be two-dimensional; a solid is three-dimensional, In more abstract studies n-dimensional spaces can be used.

2. The sizes of a plane figure or solid. The dimensions of a rectangle are its length and width; the dimensions of a rectangular parallelopiped are its length, width, and height.

3. One of the fundamental physical quantities that can be used to express other quantities. Usually, mass [M] length [L] and time [T] are chosen, Velocity, for example, has dimensions of $[L][T]^{-1}$. Force, as defined by the equation:

$$F = ma$$

where m is mass and a acceleration, has dimension $[M][L][T]^{-2}$

4. Of a matrix, the number of rows or number of columns. A matrix with 4 rows and 5 columns is 4´5 matrix.

dimensional analysis: The use of the dimensions of physical quantities to check relationships between them. For instance, Einstein's equation $E = mc^2$ can be checked. The dimensions of speed are $([L][T]^{-1})^2$, i.e. $[L]^{2}[T]^{-2}$ has dimensions of $[M][L]^{2}[T]^{-2}$. Energy also has these dimensions since it is force [M][L][T]-2 multiplied by distance [L]. Dimensional analysis is also used to obtain the units of a quantity and to suggest new equations.

directed: Having a specified positive or negative sign or a definite direction. A directed number usually has one of the signs + or - written in front of it. A directed angle is measured from one specified fine to the other. If the direction were reversed, the size of the angle would be a negative number.

direction: A property of vector quantities, usually defined in reference to a fixed origin and axes. The direction of a curve at a point is the angle from the x axis to the tangent at the point.

directional derivative: The rate of change of a function with respect to distances in a particular direction, or along a specified curve, Going from a point P(x,y,z) in the direction that makes angles a, b and c with the x, y, and z axes respectively, the directional derivative of a function f(x,y,z) is

$$df/ds = (\partial f/\partial x)\cos a + (\partial f/\partial y)\cos b + (\partial f/\partial z)\cos g$$

If there is a direction for which the directional derivative is a maximum, then this derivative is the gradient of f at point P.

direction angle: The angle between a line and one of the axes in a rectangular Cartesian coordinate system. In a plane system, it is the angle, a, that the fine makes with the positive direction of the x-axis. In three dimensions, there are three direction angles, α, β, and y, for the x, y, and z axes respectively. If two direction angles are known, the third can be calculated by the relationship:

$$\cos^2\alpha + \cos 2\beta + \cos 2\gamma = 1$$

Cosα, cosβ, and cosγ are called the direction cosines of the line, sometimes given the symbols l,m, and n. Any three numbers in the ratio l:m:n are called the direction numbers or the direction ratio of the fine. If a line joins the point $A(x_1, y_1, z_1)$ and the point $B(x_2, y_2, z_2)$ and ,the distance between A and B is D, then

$$l = (x_2 - x_1)/D$$
$$m = (y_2 - y_1)/D$$
$$n = (z_2 - z_1)/D$$

direct proof: A logical argument in which the theorem or proposition being proved is the conclusion of a step-by-step process based on a set of initial statements that are known or assumed to be true.

directrix: 1. A straight line associated with a conic, front which the shortest distance to any point on the conic maintains a constant ratio with the distance front that point to the focus.

2. A plane curve defining the base of a cone or cylinder.

discount: 1. The difference between the issue price of a stock or share and its nominal value when the issue price is less than the nominal value. Compare premium.

2. A reduction in the price of an article or commodity for payment in cash, for a large order, or for a retailer who will be selling the goods on to members of the public.

discrete: Denoting a set of events or numbers in which there are no intermediate levels, The set of integers, for examples, is discrete but the

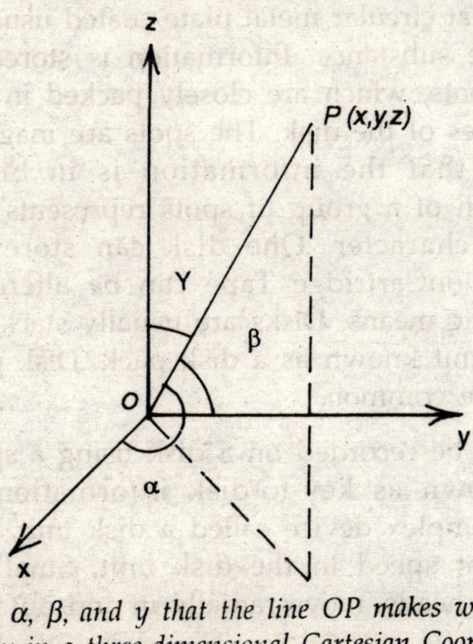

The direction angles α, β, and y that the line OP makes with the x, y, and z axes respectively in a three-dimensional Cartesian Coordinate system

set of rational numbers is not. Between any two rational numbers, no matter how close, there can always be found another rational number. The results of tossing dice form a discrete set of events, since a die has to land on one of its six faces. Putting the, shot, on the hand, does not have a discrete set of comes, since it may travel for any distance in a continuous range of lengths, .

discriminant: The expression (b^2 - 4ac) in a quadratic equation of the form $ax^2+ bx + c = 0$. If the roots of the equation arc equal, the discriminant is zero. For example in

$$x^2 - 4x+ 4 = 0$$

$b^2 - 4ac = 0$ and the only root is 2. If the discriminant is positive, the roots are different and real. If the discriminant is negative, the roots of the equation are complex numbers.

disjunction: Symbol:∨ In logic, the relationship or between two propositions or statements. Disjunction can be either inclusive or exclusive. Inclusive dis-junction is the most commonly used in mathematical-logic, and can be interpreted as one or the other or both. For two propositions P and Q, P∨ Q is false if P and Q are both false, and true in ail other cases. The more rarely used exclusive *disjunction* can be interpreted as 'either one or the other but not both'. With this definition P∨ Q is false when P and Q are both true, as well as when they are both false. The truth table ire shown, in the illustration.

disk: A device which, is widely used in computer systems to store information. It is a flat circular metal plate coated usually on both sides with a magnetizable substance. Information is stored in the form of small magnetized spots, which are closely packed in concentric tracks on the coated surfaces of the disk. The spots are magnetized in one of two directions so that the information is in binary form. The magnetization pattern of a group of spots represents a letter, digit (0-9), or some other character. One disk can store several million characters. InformationCartridge Tape can be altered or deleted as necessary by magnetic means. Disks are usually stacked on a common spindle in a single unit known as a disk pack. Disk packs storing 200 million characters are common.

Information can be recorded on a disk using a special typewriter; this method is known as key-to-disk information is fed in to a computer using a complex device called a disk unit. The disk pack is rotated at very great speed in the disk unit. Small electromagnets, known as read-write heads, move radially in and out over the surfaces of the rotating disks. They extract or record items of information at specified locations on a track, following instructions from the central processor. The time to reach a specified location is very short. This factor, together with the immense storage capacity, makes the disk unit a major backing store in a computer system.

dispersion: Any measure, of the spread of a group of numbers about their mean value. The Range standard deviation, and mean deviation are all measures dispersion.

dispersion: A measure of the extent to which data are spread about an average. The range, the difference between the largest and smallest results, is one measure. If P_r is the value below which r% of the results occur, then the range can be written as $(P_{100} - P_0)$. The interquartile range is $(P_{75} - P_{25})$. The semi-interquartile range is $(P_{75} - P_{25})/2$. The mean deviation of $X_1, X_2, ...X_n$ measures the spread about the mean X and is

$$\Sigma^n |X_i - X| / n$$

If values $X_1, X_2, ...X_k$ occur with frequencies $f_1, f_2, ...f_k$ it becomes

$$\Sigma^n f_i |X_i - X| / \Sigma f_i$$

displacement: Symbol: s The vector form of distance, measured in metres (m) and involving direction as well as magnitude,

dissipation: The removal of energy front a system to overcome some form of resistive force, without resistance there can be no dissipation. Dissipated energy normally appears as thermal energy.

distance: Symbol: d The length of the path between two points. The SI unit is the metre(m). Distance may or may not be measured in a straight fine. It is a scalar; the vector form is displacement.

distance formula: The formula for the distance between two points (x_1, y_1) and (x_2, y_2) in Cartesian coordinates. It is:

$$\sqrt{[(x_1 - x_2)^2 + (y_2 - y_1)^2]}$$

distance ratio: (velocity ratio) For a machine, the ratio of the distance moved by the effort W a given time to the distance moved by the load in the same time.

distribution function: For a random variable x, the function f(x) that is equal to the probability of each value of x occurring. If all values of x between a and b are equally likely, x has a uniform distribution in this interval and a graph of the distribution function f(x) against x is a horizontal fine. For example, the probability of the results 1 to 6 when throwing dice is a uniform distribution. Continuous random variables usually have a varying distribution function with a maximum value x and in which the probability of x decreases as x moves away from x_m.

The cumulative distri-bution function F(x) is the probability of a value less than or equal to x. For the dice example, F(x) is a step function that increases from zero to one in six equal steps, For continuous functions F(x) is often an s-shaped curve, In both cases F(x) is the are under the curve of f(x) to the left of x.

distributive: Denoting an operation that is independent of being carried out before or after another operation. For two operations * and ·; * is distributive with respect to · if a*(b·c) (a*b) ·(a*c) for all values of a, b and c. In ordinary arithmetic, multiplication is distributive with respect to addition [a(b + c) = ab + ac] and to subtraction. In set theory intersection (∩) is distributive with respect to union ∪:

$$A \cap (B \cup C) = (A \cap B) \cup (A \cap C)$$

div: (divergence) Symbol:Ñ A scalar operator that, for a three-dimensional vector function F(x, y, z), is the some of the scalar products of the unit vectors and the partial derivatives in each of the three component directions. That is:

$$\text{div } F = \nabla.F = i.\partial F/\partial x + j.\partial F/\partial y + k.\partial F/\partial z$$

In physics, div F is used to describe the excess flux leaving an, element of volume in space. This may be flow of liquid, a flow of heat in a field of varying temperature, or an electric or magnetic flux in an electric or magnetic field. If there is no source of flux within the volume, then div F = 0 and the total flux entering the volume equals the total flux leaving.

divergent sequence: A sequence in which the difference between the nth term and the one after it is constant or increases as n increases is divergent. A divergent sequence has no limit.

divergent series: A series in which the sum of all the terms after the nth term does not decrease as n increases. A divergent series, unlike a convergent series, has no sum to infinity. Compare convergent series.

dividend: 1. The number into which another number is divided to give a quotient. For example, in 16 , 3, 16 is the dividend and 3 is the divisor. 2. A share of the profits of a limited company paid to shareholders. The rate of dividend paid by a company will depend on its profits in the preceding year. It is expressed as a percentage of the nominal value of the shares.

dividers: A drawing instrument, similar to compasses, but with sharp points on both ends. Dividers are used for measuring lengths on a drawing, or for dividing straight lines.

division: Symbol: ÷ The binary operation of finding the quotient of two quantities. Division is the inverse operation to multiplication. In arithmetic, the division of two numbers is not commutative ($2 \div 3 \neq 3 \div 2$), nor associative [$(2 \div 3) \div 4 \neq 2 \div (3 \div 4)$]. The identity element for division is one only when it Comes on the right hand side.

divisor: The number by which another number is divided to give a quotient. For example, in $16 \div 3$, 16 is the dividend and 3 is the divisor.

documentation: Written instru-ctions and comments that give a full description of a computer program. The documentation describes the purposes for which the program can be used, how it operates, the exact form of the inputs and outputs, and 'how the computer must be operated. It allows the program to be amended when necessary or to be converted for use on different types of machines.

domain: A set of numbers or quantities on which a mapping is, or may be carried out. In algebra, the domain of the function f(x) is the set of values that the independent variable x can take. If, for example, f(x) represents taking the square root of x, then the domain might be defined as all the positive rational numbers.

D operator: The differential operator d/dx. The derivative df/dx of a function f(x) is often written as Df. This notation is used in solving differential equations. A second derivative, d^2f/dx^2, is written as D^2f a third derivative. d^3f/dx^3, as D^3x, and so on in some ways, the D operator can be treated like an ordinary algebraic quantity, despite the fact that it has no numerical value. For example, the differential equation

$$d\,^2y/dx^2 + 2x\,dy/dx + dy/dx + 2x = 0 \text{ or}$$

$$D^2y + 2xD + D + 2x = 0$$

can be factorized lot $(D + 2x)(D + 1) = 0$. $(D + 2x)$ then operates on the function $(D + 1)$.

double integral: The result of integrating the same function twice, first with respect to the variable, holding a second variable constant, and then with respect to the second variable, holding the first variable constant For example, if f(x, y) is a function of the variables x and y, then the double integral, first with, respect to x and then with respect to y is

$$\iint f(x, y)dydx$$

This is equivalent to summing f(x, y) over intervals of both x and y, or to finding the volume bounded by the surface representing f(x, y). The integral is not affected by the order in which the integrations are carried out if they are definite integrals. Another kind of double integral is the result of integrating twice with respect to the same variable, For example, if a cars acceleration a increases with time t in a known way, then the integral

$$\int adt$$

is the velocity (v) expressed as a function of time the double integral

$$\iint adt^2 = \int v.dt$$

where x is the distance travelled as a function of lime.

double point: A singular point on a curve at which the curve crosses itself or tangential to itself. There are several types of double point. At a node the curve crosses over itself forming a loop. In this case it has two distinct tangents. At a cusp it doubles back on itself and has only one tangent. At a tacnode two arcs of curve touch each other and have the same tangent, but unlike a cusp, the arcs continue through the singular point to form four arias. An isolated double point may also occur, This satisfies the equation of the curve but does not lie on the main arc of the curve.

drum: A metalic cylinder coated with a magnetizable substance and used in a computer system to store information. The information is stored in the form of small magnetized spots, which are closely packed in concentric tracks around the circumference of the drum. When in use the drum is rotated at high speed. Small electromagnets, called re .d-write heads, are fixed in position over each track. They extract (read) or record (write) items of information in particular locations on the track, as specified by the central processor. The time taken to obtain an item of information is extremely short. Drums are now used only for a few special applications in computing.

duodecahedron: A polyyhedron that has twelve faces. A regular *duodecahedron* has twelve congruent faces, each one being a regular pentagon.

duodecimal: Denoting or based on twelve, In a duodecimal number system there are twelve different digits instead of ten. If for example, ten and eleven were given the symbols A and B respectively, 12 would be written as, 10 and 22 as 1A. Duodecimal numbers are of little use, but some duodecimal units (1 foot = 12 inches) are still in use.

duty: A tax levied on certain kinds of transactions. Examples include taxes on importing and exporting goods and the tax levied on alcohol and tobacco products.

dynamics: The study of how objects move under the action of forces. This includes the relation between force and motion and the description of motion itself.

dyne: The former unit of force used in the c.g.s. system. It is equal to 10^{-5}N.

E

eccentricity: A measure of the shape of a conic. The eccentricity is the ratio of the distance of a point on the curve from a fixed point (the locus) to the distance from a fixed line (the directrix). For a parabola, the eccentricity is one. For a hyperbola, *it* is greater than one, and for an ellipse, it is between 0 and 1. A circle has an eccentricity of 0.

edge: A straight line where two faces of a solid meet A cube has eight edges.

efficiency: Symbol: η A measure used for processes of energy transfer; the ratio of the useful energy produced by a system or device, to the energy input. For example, the efficiency of an electric motor is the ratio of its mechanical power output to the electrical power input. There is no unit of efficiency; however efficiency is often quoted as a percentage. In practical systems some dissipation of energy always occurs and the efficiency is less than 1. For a machine, the efficiency is the force ratio divided by the distance ratio.

effort: The force applied to a machine.

elastic collision: A collision for which the restitution coefficient is equal to one, Kinetic energy is conserved during an elastic collision. In practice collisions are not perfectly elastic as some energy is transferred to internal energy of the bodies.

electronvolt: Symbol: eV A unit of energy equal to $1.602\ 1917\times 10^{-19}$ joule. It is defined as the energy required to move an electron charge across a potential difference of one volt. It is normally used only to measure energies of elementary particles, ions, or states.

element: 1. A single item that belongs to, or is a member of, a set, February', for example, is an element of the set. The number 5 is an element of the set of integers between 2 and 10. In set notation this is written as 5 ∈ {2, 3, 4, 5, 6, 7, 8, 9, 10}.

eliminant: The relationship between coefficients that results from eliminating the variable from a set of simultaneous equations For example, in the equations

$$a_1x + b_1y + c_1 = 0$$

$$a_2x + b_2y + c_2 = 0$$

$$a_3x + b_3y + c_3 = 0$$

ellipse: A conic with an eccentricity between 0 and 1. An ellipse has two foci. A line through the focus the ellipse at two vertices. The line segment between the vertices is the major axis *of* the ellipse The point on the major axis midway between the vertices is the centre of the ellipse.

A line segment through the centre perpendicular to the major axis is me minor axis. Ether of the chords of the ellipse through a focus parallel to the mirror axis is a latus rectum.

The area of an ellipse is pab, where is half the major axis and *b is* half the minor axis.

The sum Property of an ellipse is that for any point on the ellipse the sum of the distances from the point to each focus is a constant. The ellipse also has a reflection property: for a given point on the ellipse the two lines from each focus to the point make equal angles with a tangent at that point.

In Cartesian coordinates the equation:

$$x^2/a^2 + y^2/b^2 = 1$$

represents an ellipse with its centre at the origin, The major axis is on the x-axis and the minor axis on the y-axis. The major axis is 2a and the minor axis is 2b. The foci of the ellipse are at the points(+ ea, 0) and (- ea, 0), where e is the eccentricity. The directrices are the lines $x = a/e$ and $x = -a/e$. The length of the latus rectum is $2b^2/a$.

ellipsoid: A solid body or curved surface in which every plane cross section is an ellipse or a circle, An ellipsoid has three axes of symmetry. In three-dimensional Cartesian coordi-nates, the equation of an ellipsoid with its centre at the origin is:

$$x^2/a^2 + y^2/b^2 + z^2/c^2 = 1$$

where a, b, and c are the points at which it crosses the x, y, and z axes respectively. In this case the axes of symmetry are the coordinate axes. A prolate ellipsoid one generated by rotating an ellipse about its major axis. An oblate ellipsoid is generated by rotation about the minor axis.

empirical: Derived directly from experimental results or observations.

empty set: (null set) symbol: θ The set that contains no elements. For example, the set of 'natural numbers less than 0' is an empty set. This could be written as

$$\{m: m \in N; m < 0\} = \theta$$

energy: Symbol: W A property of system its capacity to do work. Energy and work have the same unit: the joule (J). It is convenient to divide energy into kinetic energy and potential energy. Names are given to many different forms of energy; the only real difference lies in the system under discussion.

enlargement: A geometrical projection that produce an image larger than, but similar to, the original shape.

entailment: In logic, the relationship that holds between two (or more) propositions when one can be deduced from the other, If conclusion C is deducible from Premisses A and B, A and B are said to entail C .

epicycle: A circle that rolls around the circumference of another, tracing out an epicycloid.

epicycloid: The plane curve traced out by a point on a circle or epicycle rolling along the outside of another fixed circle. In a two dimensional Cartesian-co-ordinate system that has a fixed circle of radius a centred at the origin and another of radius b rolling around the circumference, the epicycloid is a series of continuous arcs that move away from the first circle to a distance 2b and then return to touch it again at a cusp point where the next arc begins. The epicycloid has only one arc if a = b, two if a = b/2, and so on. If the angle between he radius from the origin to the moving point of contact between the two circles is 0, the epicycloid is defined by the parametric equations:

$$x = (a + b) \cos\theta - a\cos[(a + b)\theta/a]$$

$$y = (a + b) \sin\theta - a \sin[(a + b)\theta/a]$$

equality: Symbol: = The relationship between two quantities that have the same value or values. If two quantities are not equal, the symbol ≠ is used.

For example, $x \neq 0$ means that the variable x cannot take the value zero. When the equality is only approximate, the symbol $\approx$ is used. For example, if Δx is small compared to x then $x + (\Delta x)^2 \approx x$. When two expressions are exactly equivalent the symbol $\equiv$ is used Another kind of equality occurs when the sum of terms in an infinite series approaches a certain value as the number of terms increases. In this case the sum is asymptotically equal($\simeq$)to a number. For example, if $|x| \angle 1$, then:

$$\Sigma -x^n/n = x - x^2/2 + x^3/3 + \dots\dots\dots \simeq \log(1 + X)$$

equation: A mathematical statement that one expression is equal to another, that is two quantities joined by an equal sign. An algebraic equation contains unknown or variable quantities. It may state that two quantifies are identical for all values of the variable and in this case the identity symbol is normally used. For example

$$x^2 - 4 \equiv (x - 2)(x + 2)$$

is an identity because it is true for all values that x might have. The other kind of algebraic equation is a conditional equation, which is true only for certain values of the variables. To solve such an equation, that is, to find the values of the variables at which it is valid, it often has to be rearranged into a simpler form. In simplifying an equation, the same operation can be carried out on the expressions on both sides of the equals sign. For example,

$$2x - 3 = 4x + 2$$

can be simplified by adding 3 to both sides to give

$$2x = 4x + 5$$

then subtracting 4x from both sides to give

$$-2x = 5$$

and finally dividing both sides by 2 to obtain the solution

$$x = -5/2$$

This kind of equation is called a linear function, because the highest power of the variable is one. It could also, be written in the form $-2x - 5 = 0$. On a Cartesian coordinate graph,

$$y = -2x - 5$$

is a straight line that crosses thê x-axis at $x = 5/2$.

Performing the same operation or both sides of an equation necessarily give an equation exactly equivalent to the original. For example, starting with $x = y$ and squaring both sides gives $x^2 = y^2$ which means that $x = y$ or $x = -y$. In this case the symbol $\Rightarrow$ is used between the equations, meaning that the first implies the second, but the second does not imply the first. That is,

$$x = y \Rightarrow x^2 = y^2$$

Where the two equations are equivalent, the symbol $\Leftrightarrow$ is used, for example,

$$2x = 2 \Leftrightarrow x = 1$$

equations of motion: Equations that describe the motion of an object with constant acceleration (a). They relate the velocity v_1 of the object at the start of timing to its velocity v_2 at some later time t and the object's displacement s. They are:

$$v_2 = v_1 + at^2$$

$$s = (v_1 + v_2)t/2$$

$$s = v_1 t + at^2/2$$

$$v_2^2 = v_1^2 + 2as$$

equator: On the Earths surface, the circle formed by the plane crosssection that perpendi-cularly bisects the axis of rotation. The plane in which the circle lies is called the *equatorial plane*. A similar circle on any sphere with a defined axis is also called an equator, or equatorial circle.

equidistant: At the same distance. For example, all points on the circumference of a circle are equidistant from the centre.

equilateral: Having sides of equal length. For example, an equilateral triangle has three sides of equal length.

equilibrant: A single force that is able to balance a given set of forces and thus cause equilibrium. It is equal and opposite to the resultant of the given forces.

equilibrium: A state of constant momentum. An object is in equilibrium if:

(1) its linear momentum does not change.

(2) its angular momentum does not change. For these conditions to be met:

(1) the resultant of all outside forces acting on the object must be zero;

(2) there is no resultant turning-effect;

An object is not in equilibriurn if any of the following are true:

(1) its mass is changing;

(2) its speed is changing;

(3) its direction is changing;

(4) its rotational speed is changing.

erg: A former unit of energy used in the c.g.s. system. It is equal to 10^{-7} Joule.

error: The uncertainty in a measurement or estimate of a quantity. There are two basic types of error. *Random error* occurs in *any direction,* cannot be predicted and cannot be compensated for. It include the limitations in the accuracy of the measuring instrument and the limitation in reading it, Systematic *error* arises front faults or changes in conditions that can be corrected for. For example, if the end of a rule is worn down that 2 millimetres are missing from the scale, every measurement taken with it will be short by 2 millimetres.

escape speed: the minimum speed that an object must have in larder to escape from the surface of a planer against the gravitational attraction. The escape speed is equal to $\sqrt{(2GM/r)}$ Where G is the gravitational constant, M is the mass of the planet, and r is the radius of the planet. The concept also applies to the escape of the object front a distant orbit.

estimate: 1. A rough calculation, usually involving one or more approximations, made to give a preliminary answer to, a problem. 2. A statement indicating what the cost will be for carrying out a specified job, e.g. painting a house or repairing a car. The estimate should state the assumptions that have been made and the extent to, which the total sum is likely to vary if the assumptions are invalid.

ether: A hypothetical fluid, formerly thought to permeate all space and to be the medium through which electromagnetic waves were propagated.

Euclidan algorithm: A method of finding the highest common factor of two positive integers. The smaller number is divided inter the larger. The remainder is then divided into the smaller number, obtaining a second remainder. This second remainder is then divided into the first remainder, to, give a third remainder. This is divided into the second, and so on, until a zero remainder is obtained. The remainder preceding this is the highest common factor of the two numbers.

Euclidean geometry: A system of geometry described by the Greek mathematician Euclid in his book Elements. It is based on a number of definitions— point, line, etc—together with a number of basic assumptions. These were axioms or common notions—for example, that the whole is greater than the part and the postulates about geometric properties—for example, that a straight line is determined by two points. Using these basic ideas a large number of the theoroms were proved using formal deductive arguments. The basic assumptions of Euclid have been modified, but the system, is essentially that used today for 'pure' geometry.

One important postulate in Euclid's system is that concerned with parallel fines. Its modern form is that if a point lies outside a straight line only one straight fine can be drawn through that point parallel to the other line.

Euler characteristic: A topol-ogical property of a curve or surface, For a curve, the Euler characteristic is the number of vertices minus the number of closed continuous fine segments between. For example, any polygon has an Euler chara-cteristic *of* zero. For a surface, the Euler characteristic is equal to the number of vertices plus the number of faces minus the number of edges. For example, a cube has an Euler characteristic *of* 2, and a cylinder, a Mobius strip, and a Klein bottle have an Euler characteristic of zero.

Euler's formula: 1. The formula that relates the number *of* vertices v, faces f, and edges e in a polyhedron, that is:

$$v + f - e = 2$$

Using the theorem it can be shown that there are only five regular Polyhedroris. 2. The definition of the function $e^{i\theta}$ for any real value of θ where i is the square root of -1 as

$$e^{i\theta} = \cos\theta + i\sin\theta.$$

Any complex number $z = x + iy$ can be written in this form $x = r\cos\theta$ and $y = r\sin\theta$ are real, with r and θ representing z on an Argand diagram, Note that putting $\theta = \pi$ gives $e^{i\pi} = -1$ and $\theta = 2\pi$ gives $e^{2\pi i} = 1$

even Divisible by two. The set of even numbers is {2, 4, 6, 8, ...}.

even function: A function *f(x) of* a variable x; for which f(-x) f(x). For example, cosx and x^2 are even functions of x.

evolute: The evolute of a given curve is the locus *of* the centres of curvature of all the points on the curve. The evolute of a surface is another surface formed by the locus *of* all the centres of curvature *of* the first surface.

expansion: A quantity expressed as a sum *of* a series of terms. For instance the expression:

$$(x + 1)(x + 2)$$

can be expanded to:

$$x^2 + 3x + 2$$

Often a function can be written as an infinite series that is convergent. The function can then be approximated in any required accuracy by taking the sum of a sufficient number of terms at the beginning of the series. There are general formulae for expanding some types of expression. For example, the expansion of $(1 + x)^n$ is

$1 + nx + [n(n - 1)/2!]x^2 + [n(n - 1)(n - 2)/3 !] x^3 +$ .... where x is a variable between -1 and + 1, and n is an integer.

expected value: The value of a variable quantity that is calculated to be most likely to occur. If x can take any of the set of discrete values $\{x_1, x_2, \ldots x_n\}$ have corresponding probabilities $\{p_1, p_2, \ldots p_n\}$ then the expected value $E(X) = x_1p_1 + x_2p_2 + \ldots + x_np_n$

If x is a continuous variable with probability density function f(x) then

$$E(x) = \int_{\infty}^{\infty} xf(x)dx$$

explicit: Denoting a function that contains no dependent variables.

exponent: A number or symbol placed as a superscript after an expression to indicate the power to which it is raised. For example, x is an exponent in y^x and in $(ay + b)^x$.

The laws of exponents are used for combining exponent of numbers as follows: Multiplication:

$$x^a x^b = x^{a+b}$$

Division:

$$x^a / x^b = x^{a-b}$$

Power of a power:

$$(x^a)^b = x^{ab}$$

Negative exponent:

$$x^{-a} = 1/x^a$$

Fractional exponent:

$$x^{a/b} = {}^b\sqrt{x^a}$$

A number is raised to the power zero is equal to 1 i.e. $x^0 = 1$

exponential: A function or quantity that varies as the power of another quantity. In $y = 4^x$ y is said to vary exponentially with respect to x. The function e^x where e is the base of natural logarithms, is the exponential of x.

The infinite series

$$1 + x + x^2/2! + x^3/3! + ...+ x^n/n! + ...$$

is equal to e^x and is known as the *exponential* series. The exponential form of a complex number is

$$re^{i\theta} = r(\cos\theta + i\sin\theta)$$

exponential series: The infinite power series that is the expansion of the function e^x, namely:

$$1 + x + x^2/2! + x^3/3! + ...+ x^n/n! + ...$$

This series is convergent for all real number values of the variable x.

Replacing x by -x gives an alternating series for e^{-x}

$$1 - x + x^2/2! - x^3/3! + ...$$

Series for sinhx and coshx can be obtained by combining series for e^x and e^{-x}

expression: A combination of symbols and operations.

exterior angle: The angle formed on the outside of a plane fig. between the extension of one straight edge beyond a vertex, and the straight edge at that vertex. In a triangle the exterior angle at one vertex equals the sum of the angles on the insides of the other two vertices.

extraction: The process of finding a root of a number.

extrapolation: The process of estimating the value of a function or quantity outside a known range of values.

F

face: A flat surface on the outside of a solid figure. A cube has six identical faces.

factor: A number by which another number is divided.

factorial: The product of all the whole numbers less than or equal to a number. For example, factorial 7, written 7!, is equal to 7 × 6 × 5 ×4 × 3 × 2 × 1. Factorial zero is defined as 1.

factorization: The process of changing algebraic or numerical expressions from a sum of terms into a product. For example, the left side of the equation $4x^2 - 4x - 8 = 0$ can be factorized to $(2x + 2)(2x - 4)$ making it easy to solve for x. As the product of the two factors is 0 when either of the factors is 0, it follows that $(2x + 2) = 0$ and $(2x - 4) = 0$ will provide solutions. i.e. when x =-1 and when x = 2.

factor theorem: The condition that (x -a) is a factor of a polynomial f (x) in a variable x if and only if f (a) = 0, For example, if $f(x) = x^2 + x - 6$, f (2) 4 + 2 - 6 = 0, and f (-3) = 9 - 3 -6 =0, so the factors of f(x) arc (x 2) and (x + 3). The factor theorem is derived from the remainder theorem.

Fahrenheit degree: Symbol: °F A unit of temperature difference equal to one hundred and eightieth of the difference between the temperatures of freezing and boiling water. On the Fahrenheit scale water freezes at 32°F and boils at 212°F. To convert front a temperature on the Fahrenheit scale (T_F) to a temperature on the Celsius scale (Tc) the following formula is used: $T_F=9T_c/5 + 32$.

farad: Symbol: F The SI unit of capacitance, When the plates of a capacitor are charged by one coulomb and there is a potential difference of one volt between them, then the capacitor has a capacitance of one farad. $1 F = 1 C V^{-1}$ farad =1 coulomb per volt.

fathom: A unit of length used to, measure depth of water. It is equal to 6 feet (1.8288 m).

F distribution: The statistical distribution followed by the ratio of variances $s_1^{\ 2}/s_2^{\ 2}$ pairs of random samples, size n_1 and n_2, taken from a normal distribution it is used to compare different estimates of the same variance.

femto-: Symbol: f A prefix denoting 10^{-15}. For example, 1 femtometre (fin) 10^{-15} metre (m).

Fermat's: last theorem The theorem that the equation

$$x^n + y^n = z^n$$

where n is an integer greater than 2, can have no solution for x, y, and z. Fermat wrote in the margin of a book on equations that lie had discovered a 'truly wonderful' proof of the theorem but that the margin was too small to write it down.

fermi: A unit of length equal to 10^{-15} metre. It was formerly used in atomic and nuclear physics.

Fibonacci numbers: The infinite sequence in which successive numbers are formed by adding the, two previous numbers, that is: 1, 1, 2, 3, 5, 8, 13, 21,.........

fictitious force: A force in a system that arises because of the frame of reference of the observer. Such forces are said to be 'fictitious' because they do not actually exist they can be removed by transfer to a different frame of reference.

field: A set of numbers that can be added, multiplied, subtracted, or divided by each other (except zero) to give a result that is a member of the same set. More generally, a field can be defined as a set of entities with two operations addition and multiplication. The entities for a commutative group under addition with 0 as the identity element.

field: A region in which a particle or body exerts a force on another particle or body through space. In a gravitational field a mass is supposed to affect the properties of the surrounding space so that another mass in this region experiences a force. The region is th'ıs called a 'field of force. Electric, magnetic, and electromagnetic fields cart be similarly described. The concept of a field was introduced to explain action ai a distance.

figure: A shape formed by a combination of points, fines, curves, or surfaces.

finite set: A set that has a fixed countable number of elements. For example, the set of 'months in the year' has 12 members and is therefore a finite set.

first order differential equation: A differential equ-ation in which the highest derivative of the dependent variable is a first derivative.

floppy disk: A device that can be used to store information, similar in structure and use to a disk, but smaller and cheaper. It is a flexible plastic disk with a magnetic coating on one or both sides. It is permanently encased in a stiff envelope inside which if car be made to rotate, A read-write head operates through a slot in the envelope. The minifloppy (or minidiskette) is a smaller version.

flotation law of: An object floating in a fluid displaces its own weight of fluid. This follows from Archimedes' principle for the special case of floating objects.

flowchart: A diagram on which can be represented the major steps in a process used, say, in industry, or a problem to be investigated, or a task to be performed. A flowchart is built up from a number of boxes connected by arrowed fines. The boxes, of various shapes, have a label attached showing for example the operation or calculation to be done at each step, At a decision box a question is asked. The answer, either yes or no, determines which of two possible paths 10 take.

flywheel: A large heavy wheel used in mechanical devices. Energy is used to make the wheel rotate at high speed; the inertia of the wheel keeps the device moving at constant speed, even though there may be fluctuations in the torque. A flywheel thus acts as an energy storage device.

focal chord: A chord of a conic that passes through a focus.

focal radius: A line from the focus of a conic to a point on the conic.

focus: A point associated with a conic. The distance, between the focus and any point on the curve is in a fixed ratio (the eccentricity) to the distance between that point and line (the directrix). An ellipse has two foci. The sum of the distances to each focus is the same for ail points on the ellipse

foot Symbol The unit of length in the f.p.s. system (one third of a yard). It is equal to 0.304 8 metre.

force: Symbol: F That which tends to, change an object's momentum. Force is vector; the unit is the newton (N). In SI, this unit is so defined that:

$$F = d(mv)/dt$$

from Newton's second law

forced oscillation: The oscill-ation of a system or object at a frequency other than its natural frequency. Forced oscillation must be induced by an external periodic force.

force ratio: For a machine, the ratio of the output force to the input force. There is no unit; the ratio is, however sometimes given as a percentage. It is quite possible for force ratios fat greater than one to be obtained, Indeed may machines are designed for this so that a small effort can overcome a large load. However the efficiency cannot be greater than one and a large force ratio implies a large distance ration.

format: The arrangement of information on a printed page, on a punched card, in a computer storage device, etc., that must or should be used to meet with certain requirements.

formula: A general expression that can be applied to several different values of the quantities in question.

Foucault pendulum: A simple, pendulum consisting of a heavy bob car a long string. The period is large and the plane of vibration rotates slowly over a period of time as a result of the rotation of the Earth below it. The apparent force causing this movement is the Coriolis force.

four-colour problem: A problem in topology concerning the division of the surface of a sphere into regions. The name comes from the colouring of maps. It appears that in colouring a map it is not necessary to use more than four colours to, distinguish regions from each other.

Fourier series: A method of expanding a function by expressing it as an infinite series of periodic functions. The frequencies of the sines and cosines are increased by a constant factor with each successive term. The general mathematical form of the Fourier series is: $f(x) = a_0/2 + (a_1 \cos x + b_1 \sin x) + (a_2 \cos 2x + b_2 \sin 2x) + (a_3 \cos 3x + b_3 \sin 3x) + \ldots\ldots\ldots + (a_n \cos nx + b_n \sin nx) + \ldots$ The contains a_0, a_1, b_1, etc., called fourier coefficients are obtained by the formulae:

f.p.s system: A system of units that use, the foot, the pound and the second as the base units.

fraction: A number written as a quotient; i.e.. as one number divided by another. For example, in the fraction 2/3, 2 is known as the numerator and 3 as denominator. When both numerator and denominator are integers, the fraction is known as a simple common or vulgar fraction.

A complex fraction has another fraction as numerator or denominator for example, (2/3) / (5/7) a complex fraction. A unit fraction is a fraction that has 1 as the numerator. If the numerator of a fraction is less than the denominator, the fraction is known as a proper fraction. If net, it is an improper fraction. For example:

$$1/2 + 1/3 = 3/6 + 2/6 = 5/6$$

In multiplying fractions, the numerators, are multiplied and the denominators, multiplied. For example:

$$2/3 \times 5/7 = (2 \times 5)/ (3 \times 7) = 10/21$$

In dividing fractions, one fraction inverted thus:

$$2/3 \div 1/2 = 2/3 \times 2/1$$

frame of reference: A set of coordinate axes with which the position of any object may be specified as it changes with time. The origin of the axes and their spatial directions must be specified at every instant of time for the frame to be fully determined.

free oscillation: An oscillation ai the natural frequency of the system or object. Thus a pendulum can be forced to swing at any frequency; it will swing freely ai only one, which depends on its length and mass.

French curve: A drawing instru-ment consisting of a rigid piece of plastic or metal sheeting with curved edges. The curvature of the edges varies from being almost straight to very right curves, so that a part of the edge can be chosen to guide a pen or pencil along any desired curvature. Another instrument that serves the same purpose consist of a deformable strip of lead bar cut into short sections and surrounded by a thick layer of plastic. This is bent to from any required curvature.

frequency: Symbol f, v The number of cycles per unit time of an oscillation. The unit is the hertz (Hz). The symbol f is used for frequency, although v is often employed for the frequency of light or other electromagnetic radiation. Angular frequency (ω) is related to frequency by $\omega = 2 \pi f$

frequency curve: A smoothed frequency polygon for data which can take a continuous set of values. As the amount of data is increased and the size of class interval decreased, the frequency polygon more closely appro-ximates a smooth curve. Relative, frequency curves are smoothed relative frequency polygons.

frequency function: The fun-ction that gives the values of the frequency of each result or observation in an experiment. For a large sample that is representative of the whole population, the observed frequency function mall be the same as the probability distribution function f(x) of a population variable x.

frequency polygon: The graph obtained when the midpoints of the tops of the rectangles in a histogram with equal class intervals are joined by line segments. The area under the polygon is equal to, the total area of the rectangles.

frequency table: A table showing how,, often each type of result occurs in a sample or experiment.

friction: force opposing the relative motion of two surfaces in contact. In fact, each surface applies a force on the other in the opposite director, to the relative motion; the forces are parallel to the fine of contact. The exact causes of friction are still net fully understood. It probably results from minute surface roughness, even on apparently 'smooth' surfaces. Frictional forces do net depend on the area of contact, Presumably lubricants act by separating the surfaces. For friction between two solid surfaces, sliding friction opposes friction between two moving surfaces. It is less than the force of static, friction, which oppose,, slip between surfaces that are at rest. Rolling friction occurs when a body is rolling on a surface: here the surface in contact is constantly changing. Frictional force (F) is proportional to the force holding the bodies together (the 'normal reaction' R). The constants of proportionality are called coefficients of friction (Symbol: μ):

$$\mu = F/R$$

The laws of friction are sometimes stated:

(1) The frictional force is independent of the area of contact (for the same force holding the surfaces together.

(2) The frictional force is proportional to the force holding the surfaces together In sliding friction it is independent of the relative velocities of the surfaces.

frustrum: A geometric solid produced by two parallel planes cutting the solid, or by one plane parallel to the base.

function: Any defined proce-dure that relates one number, quantity, etc., to one or more others. In algebra, a function of a variable x is often written as f(x). If two variable quantities, x and y are related by the equation $y = x^2 + 2$, for example, then y is a function of x or $y = f(x) = x^2 + 2$. The function here means 'square the number and add two. x is the independent variable and y is called the dependent variable. The inverse function the one that expresses x in terms of y in this case would be $x = + \sqrt{(y - 2)}$, which might be written as $x = g(y)$.

A function can be regarded as a relationship between the elements of one set (the range) and those of another set (the domain), For each element of the first set there is a corresponding element of the second set into which it is 'mapped' by the function For example, the set of numbers mapped into the set {1, 8, 27, 64) by taking the cube of each element. A function may also map elements of a set into others in the same set. Within the set {all women}, there are two subsets {mothers} and {daughters}. The mapping between them is the mother of' and the inverse is the daughter of.

fundarental: The simplest way (mode) in which an object can vibrate. The fundamental frequency is the frequency of this vibration. The less simple modes of vibration are the higher harmonics; their frequencies are higher than that of the fundamental.

fundamental theorem of algebra: Every polynomial equation of the form: $a_0 z^n + a_1 z^{n-1} + a_2 z^{n-2} + \dots a_{n-1} z + a_n = 0$ in which a_0, a_1, a_2, etc are complex numbers, has at least one complex root.

fundamental theorem of calculus: The theorem used in calculating the value of a definite integral is a continuous function of x in the interval $a \le x \le b$ and if g(x) is any indefinite integral of f(x), then;

$$\int_a^b f(x)dx = [g(x)]_a^b$$

fundamental units: The units of length, mass, and time that form the basis of Most systems of units, In SI, the fundamental units are the metre, the kilogram, and the second.

furlong: A unit of length equal to one eighth osf a mile. It is equivalent to 201.168 m.

G

gamma function: The integral function

$$\Gamma(x) = \int_0^\infty t^{x-1} e^{-t} dt$$

If x is a positive integer *n*, then Γ(n) = n! If x is an integral multiple of ½ the function is a multiple of √π.

$$\Gamma(1/2) = \sqrt{\pi}$$

$$\Gamma(3/2) = \sqrt{\pi}$$

$$\Gamma(5/2) = (3/4)\sqrt{\pi}$$

$$\Gamma(7/2) = (15/8)\sqrt{\pi}$$

gauss: symbol G The unit of magnetic flux density in the c.g.s. system;. It equàl to 10^{-4} tesla.

general form: (of an equation) A formula that defines a type of relation ship between variables but does net specify values for constants. For example general form of a polynomial equation in x is

$$ax^n + bx^{n-1} + cx^{n-1} + ... = 0$$

a, b, c, etc. are constants and n is the highest integer power of x, called the degree of the polynomial. Similarly, the general form of a quadratic equation is

$$ax^2 + bx + c = 0$$

generator: A line that generates a surface; for example, in a cone, cylinder, or solid of revolution.

geodesic: A line on a surface between two points that is the shortest distance between the points. On a plane, geodesic is a straight line. On a spherical surface it is part of a great circle of the sphere.

geometric distribution: The distribution of the number of independent Bernoulli trials before a successful result is obtained, for example, the distribution of the number of titres a coin has to be tossed before a head comes up. The probability that the number of trials (x) is k is

$$P(x = k) = q^{k-1}\ p$$

The mean and variance are 1/p, and q/p^2. The moment generating function is

$$e^t p/(l - qe^t).$$

geometric sequence: A sequ-ence in which the ratio of each term to the one after it is constant, for example, (1, 3, 9, 27,...). The general formula for the nth term of a geometric sequence is $U_n = ar^n$. The ratio is called the common ratio. In the example the first term, a, is 1, the common ratio, r. is 3, and so bx^{n-1} equals 3^n. If a geometric sequence is convergent, r lies between 1 and -1 (exclusive) and the limit of the sequence is 0. That is, u^n approaches zero as n becomes infinitely large.

geometric series: A series in which the ratio of each term to the one after it is constant, for example, 1 + 2 + 4 + 8 + 16 +.... The general formula for a geometric series is $S_n = a + ar + ar^2 + ... + ar^n = a(r^n - 1)/(r - 1)$

In the example, the first term, a, is 1, the common ratio, r, is 2, and so the nth term ar^n equals 2^n. If r is greater than the series will not be convergent. If $-1 \angle r' \angle 1$ and the sum of all the terms, after the nth term can be made as small as required by making n large enough, then the series is convergent. This means that there is a finite sum even when n is infinitely large. The sum to infinity of a convergent geometric serres is a/(1 - r)

geometry: The study of solid shapes, surfaces, fines, curves, and points in space. For example, geometry deals with the measurement or calculation of angles between straight lines, the basic proper tics of circles, and the relationships between litres and points on a surface.

gill: A unit of capacity equal to one quarter of a point. A UK gill is equivalent to $1.420t \times 10^{-4}$. m^3 and a US gill is equivalent to $1.182\ 9 \times 10^{-4}\ m^3$

golden rectangle: A rectangle in which the adjacent sides are in the ratio $(1 + \sqrt{5})/2$.

golden section The division of a line of length l into two lengths a and b so that l/a = a/b, that is, a/b = (1 + √5)/2. Proportions based on the golden section are particularly pleasing to the eye and occur in many paintings, buildings, designs, etc.

Goldbach conjecture: The conje-cture that every odd number is the sum of twc prime numbers; so far, unproved.

governor: A mechanical device to control the speed of a machine. One type of simple governor consists A two loads attached to a shaft so that as the speed of rotation of the shaft increases, the loads move further outwards from the centre of rotation, while still remaining attached to the shaft. As they move outwards they operate a control that reduces the rate of fuel or energy input to the machine. As they reduce speed and move inwards they increase the fuel or energy input. Thus, on the principle of negative feedback, the speed of the machine is kept fairly constant under varying conditions of load.

grad: (gradient) symbol Ñ A vector operator that, for any function f(x,y,z), has components in the x, y, and z directions equal to the partial derivatives with respect to x, y, and z in that order. It is defined as:

$$\text{grad } f = \nabla f = i\partial f/\ \partial x + j\partial f/\ \partial y + k\partial f/\ \partial z$$

where i, j, and k are the unit vectors in the x, y, and z directions. In physics f is often used to describe the spatial variation in the magnitude of a force F in, for example, a magnetic or gravitational field. It is a vector with the direction in which the rate of change of F is a maximum, if such a maximum exists. In the Earth's gravita-tional field this would be readily towards the centre of the Earth. In a magnetic field, ∇F would point along the lines of force.

grade: Symbol: g A unit of plane angle equal to, one hundredth of a right angle, It is equivalent to, 0.9°.

gradient: 1. In rectangular Carte-sian coordinates, the rate lit which the y-coordinate of a curve or a straight lire, changes with respect to the x-coordinate. The straight line y = 2x + 4 has a gradient of + 2; y increases by two for every unit increase in x. The general equation of a straight fine is y = mx + c, where *m* is the gradient and c is a constant ((0, c) is the point at which the line cuts the y-axis, i.e. the intercept.) If *m* is negative, v decreases as x increases. For a curve, the gradient changes continuously; the gradient at a point is the gradient of the straight line that is a tangent to the curve at that point. For the curve $y = f(x)$, the gradient is the derivative dy/dx. For example, the curve $y = x^2$ has a gradient given by $dy/dx = 2x$, at any particular value of x.

graph: 1. A drawing that shows the relationship between numbers or quantities. Graphs are usually drawn with coordinate axes at right angles. For example, the heights of children of different ages can be shown by making the distance along a horizontal line represent the age in years and the distance up a vertical line represent the height in metres. A point marked on the graph ten units along and 15 units up represents a ten-year-old who is 1.5 metres fait Similarly, graphs are used to give a geometric representation of equations. The graph of $y = x^2$ is a parabola, The graph of y 3x + 10 is a straight litre. Simultaneous equations cart be involved by drawing the graphs of the equations, and finding the points where they cross. For the two equations above the graphs cross at two points: x = -2, y = 4 and x = 5, y = 25. There are various types of graph. Some, such as the histogram and the pie chart, are used to display numerical information in a form that is simple and quickly understood. Some, such as conversion graphs, are used as part of a calculation. Others, such as scatter diagrams, may be used in analysing the results of a scientific experiment.

gravitation: The concept origi-nated by Isaac Newton around 1666 to account for the apparent motion of the Moon round the Earth, the essence being a force of attraction Called gravity, between the Moon and the Earth

gravitational field: The region f space in which one body attracts other bodies as a result of their mass. To escape from this field a body has to be projected outwards with a certain speed (the escape speed). The strength of the gravitational field ai a point is given by the ratio force/mass, which is equivalent to the acceleration of free fall, g. This may be defined as GM/r^2, where G is the gravitational constant, M the mass of the object at the centre of the field and r the distance between the object and the point in question. The standard value of the acceleration of free fall at the Earth's surface is 9.8 m s^{-2}, but it varies with attitude.

gravitational mass: The mass of a body as measured by the force of attraction between masses. The value is given by Newton's law of universal gravitation. Inertial and gravitational masses appear to be equal in a uniform gravitational field.

gravity: The gravitational pull of the Earth on an object. The force of gravity on in object causes its weight

great circle: A circle, on the surface of a sphere that has the same radius as the sphere. A great circle is formed by a cross-section in any plane inter passes through the centre of the sphere.

gross: 1. Denoting a weight of goods including the weight of the containers or packing, 2. Denoting a profit calculated before deducting overhead costs, expenses, and taxes.

group: A set having certain additional properties:

(1) In a group there is a binary operation for which the elements of the set can be related in pairs, giving results that are also members of the group. For example, the set of all positive and negative numbers and zero form a group under the operation of addition. Adding any member to any other gives an element that is also a member of the group; e.g. 3 + (-2) = 1, etc.,

(2) There is an identity element for the operation — i.e. an element that combined with another — leaves it unchanged. In the example the identity element is zero: adding zero to any member leaves it unchanged; 3+ 0 = 3, etc.,

(3) For each element of the group there is, another element its inverse. Combining an element with its inverse leads to the identity element. In the example, the number +3 has an inverse -3 thus + 3 + (3)

(4) The associative law holds for the members of the group. In the example:

$$2 + (3 + 5) = (2 + 3) + 5$$

Any set of elements obeying the above rules forms a group. Note that the binary operation need not be addition, Group theory is important in many branches of mathematics for instance, in the theory of roots of equations. It is also very useful in diverse branches of science. In chemistry, group theory is used in describing symmetries of molecules to determine their energy levels and explain their spectra. In physics, certain elementary particles can be classified into mathematical groups, on the basis of their quantum numbers. Group theory has also been applied to linguistics.

group speed: If a wave motion has a phase speed that depends on wavelength, the disturbance of a progressive wave travels with a different speed than the phase speed, This is called the group speed. It is the speed with which the group of waves travels, and is given by:

$$U = c - \lambda dc/d\lambda$$

where c is the phase speed. The group speed is the one that is usually obtained by measurement If there is no dispersion of the wave motion, as for electromagnetic radiation in free space, the group and phase speeds are equal.

gyroscope: A rotating object that tends to maintain a fixed orientation in space. For example, the axis of the rotating Earth always points in the same direction towards the Pole Star . A spinning top or a cyclist are stable when moving at speed because of the gyroscopic effect. Practical applications are the navigational gyrocompass and automatic stabilizers in ships and aircraft.

H

hardware: The physical embodiment of a computer system, i.e. its electronic circuitry, disk and magnetic tape units fine printers, cabinets, etc. software.

harmonic motion: A regularly repeated sequence that can be expressed as the sum of a set of sine waves. Each component sine wave represents a possible simple harmonic motion. The complex Vibration of sound source, for instance, is a harmonic motion, as is the sound wave produced.

harmonic analysis: The use of trigonometric series to study mathematical functions.

harmonic sequence: An ordered set of numbers, the reciprocal of which have a constant difference between them; for example, {1, 1/2, 1/3, 1/ 4,.. 1/n}. The reciprocal of the terms in a harmonic sequence form an arithmetic sequence, and vice versa.

harmonic series: The sum of the terms in a harmonic sequence; for example:

$$1 + 1/2 + 1/3 + 1/4 + \ldots..$$

HCF: Highest common factor

hecto: Symbol: h A prefix denoting 10^2. For example, 1 hectometre (hm) 10^2 metres (m).

height: A vertical distance, usually upwards, from a base line or plane. For example, the perpendicular distance front the base of a triangle to the vertex opposite, and the distance between the uppermost and the base planes of a cubed, are both known as the height of the figure.

helix: A spiral-shaped space curve. A *cylindrical helix lies* on a cylinder. A conical helix lies on a cone. For example, the shape of the thread on a screw is a helix. In a straight screw, it is a cylindrical helix and in a conically tapered screw it is a cortical helix.

hemisphere: The surface boun-ded by hall of a sphere and one of its diametral planes.

henry: Symbol: H The SI unit of inductance, equal to the inductance of a closed circuit that has a magnetic flux of one weber pet ampere of current in the circuit. 1H = 1WbA·

heptagon: A plane figure with seven straight &ides. A regular heptagon has seven equal sides and seven equal angles.

Hero's formula: A formula for the area of a triangle with sides a, b, and c:

$$A = \sqrt{[s(s - a)(s - b)(s - c)]}$$

where *s* is half the perimeter.

hertz: Symbol: Hz The SI unit of frequency, defined as one cycle per second (s^{-1}). Note that the hertz is used for regularly repeated processes, such as vibration or wave motion. An irregular process, such as radioactive decay, would have units expressed as s^{-1} (per second).

heuristic: Based on trial and error, as for example sortie techniques in iterative calculations.

hexadecimal: Denoting or based on the number sixteen. A hexadecimal number is made up with sixteen different digits instead of, the ten in the decimal system. Normally' these are 0, 1, 2, 3, 4, 5, 6, 7, 8, 9, A, B, C, D, E. For example, 16 is written as 10, 21 is written as 15 (16 + 5), 59 is written as 3B [(3 × 16) +11]. Hexadecimal numbers are sometimes are used in computer systems, because they are much shorter than the long strings of binary digits that the machine normally uses. Binary numbers are easily converted into hexadecimal numbers by grouping the digits in fours.

hexagon: A plane figure with six straight sides. A regular hexagon is one with all six sides and all six angles equal, the angles ail being 120°. Congruent regular hexagons can be fitted together to cover completely a plane surface. Apart from squares and equilateral triangles, they are the only regular polygons with this property.

hexagon: A polyhedron that has six faces. For example, the cube, the cuboid, and the rhombohedron are all hexahedrons. The cube is a regular hexahedron; ail six faces are congruent squares.

histogram: A statistical graph that represents, by the height of a rectangular column, the number of times mat each class of result occurs in a sample or experiment.

homogeneous: 1. Having all the terms to, the same degree in the variables. For a homogeneous function f(x, y, z,.....) of degree *n*

$$f(kx\ ,ky,\ kz,\ ....) = k^n f(x,\ y,\ z)$$

for all values of k. For example, $X^2 + Xy + y^2$ is a homogeneous function of degree 2 and

$$(kx)^2 + kx.ky + (ky)^2 = k^2(x^2 + xy + y^2)$$

2. Describing a substance or object in which the properties do not vary with position; in particular, the density is constant throughout.

horizontal: Level, flat, and parallel to the horizon or the ground. The top of a table is a horizontal surface. On a flat page, a straight line from side to side is referred to as a horizontal line.

horse power: Symbol: HP A unit *of* power equal in 550 foot-pounds per second. It is equivalent to 746 W.

hundredweight: Symbol: cwt A unit of mass equal to 112 pounds. It is equivalent to 50.802 3 kg. In the USA a hundredweight is equal to 100 pounds, but this unit is rarely used.

hybrid computer: A computer system containing both analog and digital devices so that the properties of each can be used to the greatest advantage, For instance, a digital and an analog computer can be interconnected so that data can be transferred between them. This is achieved by means of a *hybrid* interface. Hybrid computers are designed for specific tasks and have a variety of uses, mainly in scientific and technical fields.

hydraulic press: A machine in which forces are transferred by way of pressure in a fluid. In a hydraulic press the effort FI is applied over a small area Ai and the load F, exerted over a larger area A_2. Since the pressure is the same $F_1/A_1 = F_2/A_2$. The force ratio for the machine, F_2/F_1, is A_2/A_1. Thus, in this case (and in the related hydraulic braking system and hydraulic jack) the force exerted by the user is less than the force applied; the force ratio is greater than 1. If the distance moved by the effort is si and that moved by the load is si then, since the same volume is transmitted through the system; $s_1A_1 = s_2A_2$; i.e., distance ratio, is A_2/A_1. In practice, the device is not very efficient since frictional effects are large.

hydrostatics: The study of fluids (liquids and gases) in equilibrium.

hyperbola: A conic with an eccentricity greater than 1. The hyperbola has two branches and two axes of symmetry. An axis through the foci cuts the hyperbola at two vertices. The line segment joining these vertices is the transverse axis of the hyperbola, The *conjugate* axis is a ai right angles to the transverse axis through the centre chord perpendicular to the transverse axis is a *latus rectum.*

In Cartesian coordinates the equation:

$$x^2/a^2 - y^2/b^2 = 1$$

represents a hyperbola with its centre ai the origin and the transverse axis along the x-axis. 2a is the length of the transverse axis. *2b* is the length of the conjugate axis; i.e., the distance between the vertices of a different hyperbola *with* the same asymptotes as the given one. The foci of the hyperbola are ai the points (ae,o) and where *e* is the eccentricity. The asymptotes have the equations:

$$x/a - y/b = 0,$$

$$x/a + y/b = 0.$$

The equation of the conjugate hyperbola is

$$x^2/a^2 - y^2/b^2 = 1$$

The length of the latus rectum us $2b^2/ae$. A hyperbola for which *a* and *h* are equal is a rectangular hyperbola:

$$x^2 - y^2 = a^2$$

If a rectangular hyperbola is rotated so that the x and y axes are asymptotes, then its equation is

$$xy = k$$

where *k* is a constant.

hyperbolic functions: A set of functions that have properties similar in sortie ways to the trigonometric functions, called the hyperbolic sine, hyperbolic cosine, etc. They are related to the hyperbola in the way that the trigonometric functions (circular functions) are related to the circle.

The hyperbolic sine (sinh) of an angle a is defined as:

$$\sin\alpha = \tfrac{1}{2}(e^a - e^{-a})$$

The hyperbolic cosine (cosh) of an angle a is defined as:

$$\cosh\alpha = \tfrac{1}{2}(e^a + e^{-a})$$

The *hyperbolic tangent* (tanh) of an angle a is defined as:

$$\tanh\alpha = \sinh\alpha/\cosh\alpha =$$

$$e^a - e^{-a})/e^a + e^{-a})$$

Hyperbolic secant (sech), hyperbolic cosecant (cosech), and hyperbolic cotangent (coth) are defined as the reciprocal of cosh, sinh, and tanh respectively. Some of the fundamental relationships between hyperbolic functions are:

$$\sinh(-\alpha) = \sinh\alpha$$

$$\cosh(-\alpha) = +\cosh\alpha$$

$$\cosh^2\alpha - \sin^2\alpha = 1$$

$$\text{sech}^2\alpha + \tanh^2\alpha = 1$$

$$\coth^2\alpha - \text{cosech}^2\alpha = 1$$

hyperboloid: A surface gene-rated by rotating a hyperbola about one of its axes of symmetry. Rotation about the conjugate axis gives a *hyperboloid* of one sheet. Rotation about the transverse axis gives a hyperboloid of two sheets.

hypotenuse: The side opposite the right angle in a right-angled triangle. The ratios of the hypotenuse length to the lengths of the other sides are used in trigonometry to define the sine and cosine functions of angles.

hypothesis: A statement, theory, or formula that has yet to be proved but is assumed to be true for the purposes of the argument.

hypothesis test: A rule for deciding whether an assumption about the distribution of a random variable should be accepted or rejected, using a sample from the distribution. The assumption is called the mail hypothesis, written H_0 and it is tested against some alternative hypothesis, H_1. For example, when a coin is tossed H_0 can be P (heads) = ½ and H_1 that P (heads) > ½ A statistic is computed from the sample data. If it falls in the critical region, *where the* value of the statistic is significantly different front that expected under H_0, H_0 is rejected in favour of HI. Otherwise H_0 is accepted. A type 1 error occurs if H_0 is rejected when it should be accepted. A type II error occurs if it is accepted when it should be rejected. The significance level of the test, a is the maximum probability with which a type 1 error can be risked. For example, a = 1% means H_0 is wrongly rejected in one case out of 100.

I

irosahedron: A polihedron that.has twenty faces. A regular icosahedron has twenty congruent face each the one an equalaterl triangle.

identity element: An element of a set that, combined with another element, leaves it unchanged.

identity set: A set consisting of the same elements as another. For example, the set of natural numbers greater than 2 the set of integers greater that 2 are identity sets.

image: The lesuit of a geometrical transformation or a mapping. For example, in geometry, when a point or a set of points are transformed into another by reflection in a fine, the rellected figure is called the image. Similarly, the result of a rotation or a projection is called the image. The algebraic equivalent of this occurs when a function $f(x)$ acts on a set A of values of x to produce an image set B.

imaginary number: A multiple of i, the square root of minus one. The use of imaginary numbers is needed to solve equations such as $x^2 + 2 = 0$, for which the solutions are $x = -i \sqrt{2}$ and $x = \sqrt{2}$.

imperial units: The system of memurement based on the yard and the pound, formerly used in the UK. The f.p.s. system was a scientific system based on Imperia, units.

P.Q	*P→Q*
TT	*T*
TF	*F*
FT	*T*
FF	*T*

implication

implication: 1. (material implication conditional) Symbol: → or ⊃ In logic, the relatlonship if ... then ... between statements. Strictly reflects its ordinary language interpretation (if...then...) much then much less than conjunction, disjunction, and negative do theirs. Formally, $P \to Q$ is equivalent to either not P or Q ($\sim P$) hence $P \vee Q$ is false only when P is true and Q is false. Thus iog calvsi, eaking, if 'pigs car fly' is substituted for P and 'grass is green' for Q then 'if pigs cari fly then grass is green' is true. The truth table definition of implicationis given in the illustration. 2. In algebra, the symbol ⇒ is used between two equations when the first implies the second. For example:

$$x = y \Rightarrow x^2 = y^2$$

implicit: Denoting a function that contains two or more variables that are net independent of each other. An implicit function of x and y is one of the form $f(x,y) - 0$, for example,

$$x^2 + y^2 - 4 = 0$$

Sometimes an *explicit function*, that is, one expressed in terms of an independent variable, can be derived from an implicit function. For example,

$$y + x^2 - 1 = 0$$

can be written as

$$y = 1 - x^2$$

where y is an explicit function of x.

impulse: (impulsive force) A force acting for a very short time, as in a collision. If the force (F) is constant the impulse is $F\delta t$, δt being the time period. If theforce is variable the impulse is the integral of this over the short time period. An impulse is equal to the change of momenturn that it produces

inch: Symbol in or ″ A unit of length equal to one twelfth of a foot. It is equivalenl to 0.0254 m.

incircle: (inscrribed circte) A circle draw ninside a figure touching all the slides.

inclined plane: A type of machine. Effectively a plane at an angle, it can be used to raise a weight vertically by movement up an incline. Both distance ratio and force ratio depend on the angle of inclination (θ) l and equal 1/sin θ. The efficiency can be fairly high if friction is low. The screw and the wedge are both examples of inclined planes.

increment: A small difference in a variable. For example, x might change by an increment Δx from the value x. to the value x_2. In calculus, infinitely small increments are used.

indefinite integral: The general integration of a function, $f(x)$, of a single variable, x, without specifying the interval of x to which it applies. For example, if $f(x) = x^2$, the indefinite integral

$$\int f(x)dx = \int x^2 dx = x^3/3 + C$$

where C is an unknown constant that depends on the interval.

indeterminate equation: An equation that has an infinite number of solution. For example,

$$x + 2y = 3$$

is indeterminate. because an infinitenumber of values of x and y will satisfy it. An indeterminate equation in which the variables can take only integer va1ue, is called a Diophantine equation and ibas an infinite but denumer-able set of solutions.

indeterminate form: An expression that can have no quantitative meaning; for example 0/0.

index: A number that indicates a charactcristic or function in a mathematical expression. For example, in y^4, the exponent, 4, is also known as the index. Similarly in ^{3}27 and $\log_{10}x$, the number 3 and 10 respectively are called indices.

indirect proof: (reductio ad absurdum) Alogical argu-ment in winch a proposition or statement is proved hy howing that its negation or denial leads to a contradiction.

induction: 1. mathemalical induction) A of proving mathernatical theorems, used particularly for series sums. For instance, it is possible to show that the series 1 + 2 + 3 + 4 + ... has a sum to n of $n(n+1)/2$. First we show that if is true for n term is must also be true for $(n + 1)$ terms to the formula

$$Sn = n(n + 1)/2$$

If the formula is correct, the sum to $(n + 1)$ terms is obtained by adding $(n + 1)$ terms is obtained by adding $(n + 1)$, i.e.:

$$S_{n+1} = n(n+1)/2 = (n+1)(n+2)/2$$
$$S_{n+1} = (n+1)(n+2)/2$$

This agrees with the result obtained by replacing n in the general formula by $(n + 1)$,

$$S_{n+1} = (n + 1)(n + 1 + 1)/2$$
$$S_{n+1}\ (n + 1)(n + 2)/2$$

Thus, the formula is true for $(n + 1)$ terms if it is true for n terms. Therefore, if it is irue for the sum to one term $(n = 1)$, a must be true for the sum to two terms $(n + 1)$. Similarly, if true for two terms true for three terms, and so on through all values of n. It is easy to show that it is true for one term:

$$S_n = + 1(1 + 1)/2$$
$$S_n = 1$$

which is the first term in the series. Hence the theorem is true for all integer values of n.

2. In logic, a form of reasoning from individual cases to general ones, or from observed instances to unobserved ones. Inductive arguments can be of the form: F_1 is A, F_2 is A... F_n is A, therefore all *Fs* are *A* ('this swan has wings ...that the swan has wings or: all *Fs* observed so far are A. Unlike deduction, asserting the premisses while denying the conclusion in an induction does not lead to a contradiction.

inelastic collision: A collision for which the restitution coefficient is less than one In effect, the relative velocity after the collision is less than that before; the kinetic energy of the bodies is not conserved in the collision even though the system may be closed. Some of the kinetic energy is converted into internal energy.

inequality: A relationship between two expressions that are not equal, often written in the form of an equation but with the symbols > or < meaning 'is greater than' and 'is less than'. For example, if $x < 4$ then $x^2 < 16$. If $y^2 > 25$, then $y > 5$ or $y < -5$. If the end values are included the symbols ≥ —is greater than or equal to — and ≤ —is less than or equal to— are used. When one quantity is very much smaller than another it is shown by < < or > >. Fore example, if x is a large number $x > > 1/\ x < < x$.

inertia: An inherent property of matter implied by Newton's first law of motion: inertia is the tendency of a body to resist change in its motion.

inertial mass: The mass of an object as measured by the property of inertia. It is equal to the ratio force/acceleration when the object is accelerated by a constant force. In a uniform gravitational field, it appears to be equal to gravitation mass all objects have the same gravitational accelera-tion at the same place.

inertial system: A frame of reference in which an observer sees an object that is free of all external forces to be moving at constant velocity. The observer is called an inertial observer. Any frame of reference that moves with constant velocity and without rotation relative to an inertial frame is also an inertial frame. Newton's laws of motion are valid in any inertial frame (but not in an accelerated frame), and the laws are therefore independent of the velocity of an inertial observer.

inference: The process of reaching a conclusion from a set of premisses in a logical argument. An inference may be deductive or inductive.

infinite set: A set in which the number of elements is infinite. For example, the set of 'positive integers', z {1, 2, 3, 4....}, is infinite but the set of 'positive integers' less than 20' is a *finife* set. Another example of an infinite set is the number of circles in a particular plane.

infinitesimal: Infinitely small but is equal to zero. Infinitesimal change or differences are made use of in calculus.

infinity: Symbol ∞ The value of a quantity that increases without limit. For example, if $y = 1/x$, then y becomes infinitely large, or approaches infinity, as x approaches 0. An infinitely large negative quantity is denoted by $-\infty$ and an infinitely large positive quantity by +¥. If x is positive, $(y = -(1/x))$ tends to $-\infty$ as x tends to 0.

information theory: The branch of probability theory that deals with uncertainty, accuracy, and information content in the transmission of massages. It can be applied to any system of communication including electrical signals and human speech. Random signals (noise) are often added to a message during the transmission process altering the signal received from that sent. Information theory is used to workout the probability that a particular signal received is the same as the signal sent. Redundancy, for example simply repeating a message, is needed to overcome the limitations of the system. Redundancv can also take the form of a more complex checking process. In transmitting a sequence of numbers, their sum might also be transmitted so that the receiver will know that there is an error when the sum does not correspond to the rest of the message. The sum itself gives no extra information since, if the other numbers are correctly received, the sum can easily be calculated. The statistics of choosing a message out of all possible messages determines the amount of information contained in it. Informations measured in bits. If one out of two possible signals are sent then the information content is one bit. A choice to one out of four possible signals contains more information, although the signal itself might be the same.

input: 1. The signal or other form of information that is applied (fed in) to an electrical device, machine, etc. The input to a computer is the data and programmed instructions that a user communicates to the machine. An input device accepts computer input in some appropriate form and converts the information into a code of electrical pulses. The pulses are then transmitted to the central process of the computer. There are various input devices, including paper tape readers and card readers.

input/output (I/O): The equipment and operators used to communicate with a computer, and the informa-tion passed in or out during the communication. Input/output devices include those used only for input or for output of information and those, such as visual display units used for both input and output.

inscribed: Describing a geo-metric figure that is drawn inside another geometrical figure.

instantaneous value: The value of a varying quantity at a particular instant in time.

integers: Symbol: z The set of whole numbers (...., 2, 1, 0, 1, 2,.....) used for counting. It includes zero and the negative whole numbers.

integrand: A function that is to be integrated. For example, in the integral of $f(x).dx$, $f(x)$ is the integrand.

integrating factor: A multiplier used to simplify and solve differential equations, usually given the symbol ξ. For example, $xdy - ydx = 2x^3dx$ may be multiplied by $\xi(x) = 1/x^2$ to give the standard form:

$d(y/x) = (x.dy - y.dx)/x^2$ which has $= (x.dy - y.dx)x^2 = 2x\ dx$, winch has the solution $y/x = x^2 + c$ where c is a constant.

integration: The continuous summing of change in function, f(x) over an interval of the variable x. It is the inverse process of differentiation in calculus, and its result is known as the integral of f(x) with respect to x. For example, the total distance a car travels along a road in time interval t_1 to t_2 is the integral of its velocity v over this interval is written

$$v.dt$$

Because this has defined limit t_1 and t_2 it is called a definite integral. More generally

$$v.dt$$

Because this has defined limits, t_1 and t_2 it is called a definite integral. More generally

$$x = \int v.dt$$

is the area between the curve and the x-axis, between the values x_1 and x_2. it can be considered as the sum of number of column area of width Δx and heights given by f(x). As Δx approaches zero. The number of column increases infinitely and the sum of the column areas approaches the area under the curve.

interaction: Any mutual action between particles, systems, etc. Examples of interactions include the mutual forces of attraction between masses and the attractive or repulsive forces between electric charges.

intercept: A part of a fine or plane cut off by another fine or plane.

interest: The amount of money paid each year at a stated rate on a borrowed capital, or the amount received each year at a stated rate on loaned or invested capital. The interest rate is usually expressed as a percentage per annum.

interface: A shared boundary. It is the area(s) or place(s) at which two devices or two systems meet and interact. There is a simple interface between an electric plug and socket. A far more complicated interface of electronic circuits provides the connection between the central processor of a computer and each of its peripheral units. Manma-chine interface refers to the interaction between people and machines, including computers.

interior angle: An angle formed on the inside of a plane figure by two of its straight sides. For example, there are the interior angles in a triangle, which add up to 180°.

interpolation: The process of estimatıng the value of a function from known values on either side of it. For example the speed of an engine, controlled by a lever increases from 40 to 50 revolution per second when the lever is pulled down by four centimetres, one can interpolate from this information assume that moving it two centimetres will give 45 revolutions per second. This is the simplest method of interpolation, called *linear interpolation.* If known values of one variable y are plotted against the other variable x an estimate of an unknown value of y can be made by drawing a straight line between two nearest known values.

The mathematical formula of linear interpolation is:

$$y_3=y_1+(x_3-x_1)\ (y_2-y_1)/(x_2-x_1)$$

y_3 is the unknown y_2 and y_1 (at x_2 and x_1) are nearest known values, between which the interpolation is made. If the graph of y against x is a smooth curve, and the interval between y_1 and y_2 is small; linear interpolation can give a good approximation to the true value, but $(y_2 - y_1)$ is large, it is less likely that will fit sufficiently well known to a straight line between y_1 and y_2. A possible source of error occurs when y is known at regular intervals, but oscillates with a period shorter than this interval.

interquartile range: A measure of dispersion given by $(P_{75} - P_{25})$ where P_{75} is the upper quartile and P_{25} the lower quartile. The semi-interquartile range is $\frac{1}{2}(P_{75} - P_{25})$

intersection: 1. The point at which two or more lines cross each other, or a points that two or more geometrical figures have in common.

2. set theory, the set formed by the elements common to two or more sets. For example, if set A is black four legged animals} and set B is {sheep} thee the section of A and B, written $A \cap B$, is {black sheep}. This can be represented on a Venn diagram by the intersection of circles, one representing A and the other, B.

interval: A set of numbers, or points in a coordination system, defined as all the values between two end points.

inverse: The matrix that gives as a product with another, the identity matrix I. Given a square matrix A with inverse A^{-1}, then $AA^{-1} = I$. The inverse is only defined for square matrices in which the determinant is not zero. A matrix equation $Y = AX$ can be multiplied by A^{-1}, to obtain $X = A^{-1}Y$. This is used to solve simultaneous equations, where the matrices X and Y represent the coefficients and constants in the equation. For example, the equations

$$x + 3y = 5$$
$$2x + 4y = 6$$

can be represented by a matrix equation. The solution is $x = -1$ and $y = -2$. When there are three or more simultaneous equations with three or more variables, this technique is easier than other methods, because there are relatively easy techniques available for finding the inverses of matrices.

inverse element: An element of a set that, combined with another element gives the identity element.

inverse function: The reverse mapping of the set B into the set A when the function that maps A into B has already been defined.

inverse hyperbolic functions: The inverse functions of hyperbolic cosine, hyper-bolic cosine, hyperbolic tangent etc., defined in an analogous way to the inverse trigonometric functions. For instance, the inverse hyperbolic sine of a variable x, written arc sinh$x$ or $\sinh^{-1}x$ is the angle of which x is the hyperbolic sine. Similarly, the other inverse hyperbolic functions are:

inverse hyperbolic cosine of x (written arc cosh$x$ or cash^{-1} x)

inverse hyperbolic tangent of x (written arc tanh$x$ or $\tanh^{-1} x$)

inverse hyperbolic cotangent of x (written arc coth x or cothx)

inverse hyperbolic secant of x (written are sech$x$ or sech^{-1} x)

inverse hyperbolic cosecant of x (written arc cosech$x$ or cosech^{-1}x).

inverse ratio: A reciprocal ratio, For example, the inverse ratio of x to y is the ratio of $1/x$ to $1/y$.

inverse square law: A physical law in which an effect varies inversely as the square of the distance from the source producing the effect an example Newton's law of universal gravitation.

involute: The involute of a curve is a second curve that would be obtained by unwinding a taut string wrapped around the first curve. The involute is the curve traced out by the end of the string.

irrational number: A number that cannot be written as an exact ratio of two integers. For example, the square root of three ($\sqrt{3}$ = 1.732 0508...) can be calculated to any desired degree of accuracy, but can only be exactly defined as the largest number of which the square does not exceed three. Such a number is called a surd. Another type of irrational number is the *transcendental number*, which arises not as the result of a simple algebraic relationship, but as a basic property of mathematics. For example, π and e are transcendental numbers, which arise in geometry and calculus respectively. Trans-cendental numbers are defined as numbers that are not the roots of any algebraic equation with rational coefficients.

isolated point: A point that satisfies the equation of a curve but is non on the main arc of the curve. For example the equation $y^2 (x^2 - 4) = x^4$ has a solution at $x = 0$, and $y = 0$ but there is no real solution at any point near the origin so the origin is an isolated point.

isometry: A transformation in which the distances between the points remain constant.

isomorphic: Having a one-to-one correspondence. In topology, two sets of points that are isomorphic are topologically equivalent.

isomorphism: A one-to-one correspondence between two sets. Each element of the first set can be related to an element in the second set by an operation. For instance, multiplication by a constant integer relates one set of integers to another set of integers. A relationship that does not conform in this, for example, taking the square root will form a *non-isomorphism*.

J

job: A unit of work submitted to computer. It usually includes several programs. The information necessary to run a job is input in the form of a short program written in the job-control language (JCL) of the computer. The JCL is interpreted by the operating system and is used to identify the job and describe its requirements to the operating system.

joule Symbol: J The SI unit of energy and work, equal to the work done when the point of application of a force of one newton moves one metre in the direction of action of the force. 1J = 1 N m. The joule is the unit of all forms of energy.

K

kelvin Symbol: K The SI base unit of thermodynamic temperature. It is defined as the fraction I/273.16 of the thermodynamic temperature of the triple point of water. Zero kelvin (10K) is absolute zero. One kelvin is t the same as one degree on the Celsius scale of temperature.

Kendall's method: A method of measuring the degree of association between two different ways of ranking *n* objects, using two variables (*x* and *y*), which give data $(x_1,y_1)...,(x_m y_n)$. The objects are ranked using first the *x*s and then the *y*'s. For each of the $2n(n-1)/2$ pairs of objects a score is assigned. If the rank of the *j*th object is greater (or less) than that of the *k*th, regardless of whether the *x*'s or *y*'s are used, the score is plus one. If the rank of the *j*th is less than that of the *k*th using one variable but greater using the other, the score is minus one. Kendall's coefficient of rank correlation τ = (sum of scores)/½n − 1). The closer π is to one, the greater the degree of association between the rankings.

Kepler's laws: Laws of planetary motion deduced in about 1610 by Johannes Kepler using astronomical observations made by Tycho Brahe:.

(1) Each planet moves in an elliptical orbit with the Sun at one locus of the ellipse.

(2) The fine between each planet and the Sun sweeps out equal areas in equal times.

(3) The square of the period of each planet is proportional to the cube of the semi-major axis of the ellipse.

Application of the third law to the orbit of the Moon about the Earth gave support to Newton's theory of gravitation.

kilo: Symbol k A prefix denoting 10^3. For example, 1 kilometre (km) 10^3 metres (m).

kilogram: (kilogramme) sym-bol kg The SI base unit of mass, equal to the mass of the international prototype of the kilogram, which is a piece of platinum-iridium kept at Serres in France.

kilowatt-hour: Symbol kwh A unit of energy; usually electrical, equal to the energy transferred by one kilowatt of power in one tour. It is the same as the Board of Trade unit and has a value of 3.6×10^6 joules.

kinematics: The study of the motion of objects without consideration of its cause.

kinetic energy: Symbol T The work that an object can do because of its motion. For an object of mass m moving with velocity v, the kinetic energy is given by $mv^2/2$. This gives the work the object would do in coming to rest. The rotational kinetic energy of an object of moment of inertia I and angular velocity ω is given by $lw^2/2$.

kite: A plane figure with four sides, and two pairs of adjacent sides equal. Two of the angles in a kite are opposite and equal. Its diagonals cross perpendicu-larly, one of them (the shorter one) being bisected by the other. The area of a kite is equal to the product of its diagonal lengths. In the special case in which the two diagonals have equal lengths, the kite is a rhombus.

Klein bottle: A curved surface with the unique topological property that it has only one surface, no edges, and no inside and outside. It can be thought of as formed by taking a length of flexible stretchy tubing, cutting a hole in the side through which one end can fit exactly, passing an end of the tube through this hole, and then joining it to the other end front the inside. Starting at any point on the surface a continuous line can be drawn along it to any other point without crossing an edge.

knot: In topology, a curve formed by looping and interlacing a string and then joining the ends. The mathematical theory of knots is a branch of topology.

knot: A unit of velocity equal to one nautical mile per hour. It is equal to 0.514 ms^{-1}.

Konigsberg bridge problem: A classical problem in topology. The river in the Prussian city of Konigsberg divided into two branches and was crossed by seven bridges in a certain arrange-ment. The problem was to show that it is impossible to walk in a continuous path across all the bridges and cross each one only once. The problem was solved by Euler in the eighteenth century, by replacing the arrangement by an equivalent one of lines and vertices. He showed that a network like this can be traversed in a single path if and only if there are fewer than three vertices at which an odd number of line segments meet. In this case there are four.

L

latitude: The distance of a point on the Earth's surface from the equator, measured as the angle in degrees between a plane through the equator the equatorial plane) and the line from the point to the centre of the Earth. A point en the equator has a latitude of 0° and the North Pole has a latitude of 90°.

law of the mean: (mean value theorem) The rule in differential calculus that, if $f(x)$ is continuous in the interval $a \leq x \leq b$, and the derivative $f'(x)$, exists everywhere in this interval, then there is at si one value of x (x_0) between a and b for which:

$$[f(b) - f(a)]jl(b - a) \ f(x_0)$$

Geometrically this means that if a straight line is drawn between two points, $(a, f(a))$ and $(bf(b)$, on a continuous curve, then there is at learnt one point between these where the tangent to the curve is parallel to the line. This law is derived from Rolle's theorem.

Laws of Thought: Three laws of logic that arc traditionally considered– as are other logical rules to exemplify something fundamental about the way we think; that is, it is not arbitrary hat we say certain forms of reasoning are correct. On the contrary, it would be impossible to think other-wise. 1. The law of contradic-tion. It is not the case that something can both be true and not be true: symbolically

$$\sim (p \wedge \sim p)$$

2. The law of excluded Middle. Something must either be true or not be true symbolically

$$p \vee p$$

3. The law of identity. If something is true, then it is true: symbolically

$$p \rightarrow p$$

league: A unit of length equal to 3 miles. It is equivalent to 4828.032 m

least squares method: A method of fitting a regression line to a set of data. If the data are points (x_1,y_1), ... $(x_m y_n)$ the corresponding points are found using the linear equation $y = ax + b$. The least squares line minimizes $[(y_1 - y_2)^2 + (y_2 - y_1)^2 + ... + (y_n - y_n)^2]$. It is found by solving the normal equations

$$\Sigma y = an + \Sigma x$$

$$\Sigma xy - a\Sigma x + b\Sigma x^2$$

for a and b. The technique is extended for regression quadratics, cubics, etc.

Legendre polynomial: A series of functions that arises as solutions to Laplace's equation in spherical polar coordinates. They form an infinite series.

lemma: A theorem that has already been proved and is used as a basic assumption (axiom or premiss) in another proof.

length: The distance along a line, plane figure, or solid. In a rectangle the greater of the two dimensions is usually called the length and the smaller, the breadth.

lever: A class of machine; a rigid object able to turn around some point (pivot or fulcrum). The force ratio and the distance ratio depend on the relative positions of the pivot, the point where the user exerts the effort, and the point where the lever applies force to the load. There are three types (orders) of lever.

First order, in which the fulcrum is between the load and the effort. An example is a crowbar.

Second order, in which the load is between the effort and the fulcrum. An example is a wheelbarrow.

Third order, in which the effort is between the load and the fulcrum. An example is a pair of sugar tongs.

Levers can have high efficiency; the main energy losses are by friction at the pivot, and by bending of the lever itself.

library programs: Collections of computer programs that have been purchased contributed by users, or supplied by the computer manufacturers for the use of the computing community. Libraries that have been purchased or supplied by manufacturers usually have a theme, such as engineering or mathematics.

light-year Symbol: ly A unit of distance used in astronomy, defined as the distance that light travels through space in one year. It is approximately equal to 9.460 5 $\times 10^{15}$ metres.

limit: In general, the value, approached by a function as the independent variable approaches some value. The idea of a limits the basis of the branch of mathematics known as analysis. There are several example of the use of limits. (1) The limit of a function is the value it approaches as the independent variable. For instance, the function $x/(x + 3)$ for positive value of x is less than 1. As x increases it approaches 1 – the value approached as x becomes infinite. This is written

$$\lim_{x \to \infty} x/(x + 3) = 1$$

stated as 'the limit of $x/(x + 3)$ as x approaches (or lends to) infinity is 1'.1 is the limiting value of the function. (2) The limit of a convergent sequences is the limit of the nth term as n approaches infinity. (3) The limit of a convergent series is the limit of the sum of n terms as n approaches infinity. (4) A derivative of a function $f(x)$ is the limit of $[f(x + \sigma x)\ f(x)]/\sigma x$ as x approaches zero. (5) A definite integral is the limit of a finite sum of terms $y\sigma x$ as fx approaches zero.

line: A join between two points in space or on a surface. A line has length but no breadth, that is, it has only one dimension. A straight fine is the shortest distance between two points on a flat surface.

linear equation: An equation in which the highest power of an unknown variable is one. The general form of a linear equation is $mx + c = 0$

where m and c are constants. On a Cartesian coordinate graph

$$y = mx + c$$

is a straight line that has a gradient m and crosses the y-axis at $y = c$. The equation

$$x + 4y^2 = 4$$

is linear in x but not in y.

linear programming: The process of finding maximum or minimum values of a linear function under limiting conditions or constraints. For example the function $x - 3y$ might be minimized subject to the constraints that $x + y, \leq 10$, $x = y$, $x > 0$ and $y \geq 0$. The constraints can be shown as the area on a Cartesian coordinate graph bounded by the lines $x + y = 10$; $x = y$, $x = 0$, and $y = 0$. The minimum value for $x - 3y$ is chosen from points within this area. A series of parallel lines $x - 3y = k$ arc drawn for different values of k. The line $k = -9$ just reaches the constraint area at the point (10,0), Lower values are outside it, and so $x = 10$, $y = 0$ gives the minimum value of $x - 3y$ within the constraints. Linear programming is used of find the best possible combination of two, or more variable quantities that determine the value of another quantity. In most applications for example, finding the best combination of quantities of each product from a factory to give the maximum profit, there are many variables and constraints. Linear functions with large number of variables and constraints are maximized or minimized by computer techniques that are similar in principle his graphical technique for two, variables.

line integral: (contour integral, curvilinear integral) The integration of a function along a particular path, C, which may be a segment of straight line, a portion of a space curve, or connected segments of several curves. The function is integrated with respect to the position vector $r = xi + yj + zk$, which denote, the position of each point P(x, y, z) on a curve C.

For example, the direction and magnitude of a force vector F acting or a particle may depend on the particle's position in a gravitational field or a magnetic field. The work done by the force in moving the particle over a distance dr is Fdr. The total work done in moving the particle along a particular path from, point P_1 to point P_2 is the line integral

$$F.\mathrm{d}r = F_x\mathrm{d}x + F_y\mathrm{d}y + F_z\mathrm{d}z$$

line printer: An output device of a computer system that prints characters (letters, numbers, punctuation marks, etc.) on paper a complete line at a time and can therefore operate very rapidly 1000 lines per minute is a typical speed. The barrel (or *drum*) printer is widely used. It has a set of printing character, embossed around the circumference of a barrel at each character position across the page. The paper is continuous, with a line of perforations between each sheet and sprocket belles along each side to control its movement.

Lissajous figures: Patterns obtained by combining simple harmonic motions in different directions. For example, an object moving in a plane so that two movements of the motion at right angles are simple harmonic motions, traces out a Lissajous figure. If the components have the same frequency and amplitude and are in time the motion is a straight line. If they are out of phase by 90°, it is a circle. Other phase differences produce ellipses. If the frequencies of the components differ, more complex pattern are formed. Lissajous figures can be demonstrated with one oscilloscope, deflecting the spot with one oscillating signal along one axis and with another signal along the other axis.

litre: Symbol 1 A unit of volume defined as 10^{-3} metre^3. The name is not recommended for precise measurements. Formerly, the litre was defined as the volume, of one kilogram of pure water at 4°C and standard pressure. On this definition, 11=1000.028 cm^3.

local maximum: (relative maximum) A value of a function f(x), that is greater than for the adjacent values of x, but is not the greatest of all values of x,

local minimum: relative minimum) A value of a function f(x) hall is less than for the adjacent values of x but is not the lowest of ail values of x

locus of points: A set of points, often defined by an equation relating coordi-nates. For example, in rectangular Cartesian coordi-nates, the locus of points on a line through the origin at 45° to the x-axis and the y-axis is defined by the equation $x = y$; the lines said to be the locus of the equation. Any geometrical figure an interval on a line, a circle in a plane, a cube–is a locus of points.

logarithm: A number expressed as the exponent of another number. Any number x can be written in the form $x = z$, y is then the logarithm, to the base a of x. For example, the logarithm to the base ten of 100 ($\log_{10}100$) is two since $100 = 10^2$. Logarithms that have a base of ten are known as common logarithms (or *Briggsian* logarithms). They are used to carry out multiplication and division calculations because numbers can be multiplied by adding their logarithms. In general $p \times q$ can written as is, $a^c \times a^d = a(c + d)$, $p = a^c$ and $q = a^d$. Both logarithms and antilogarithms (the inverse function) are available in the form of printed tables, 4.91 × 5.12 would be calculated as follows: $\log_{10}$ 4.91 is 0.6911 (from tables) and $\log_{10}$ 5.12 is 0,7093. Therefore, 4.91 × 5.12 is given by antilog (0.691 + 0.7093) which is 25.14 (from antilog tables). Similarly, division can be carried out by subtraction of logarithms and the nth root of a number (x) is the antilogarithm of $(\log x)/n$.

For numbers between 0 and 1 common logarithm is negative, For example, $\log_{10} 0.01 = -2$. The common logarithm of any positive real number be written in the form n where x is between 1 and 10 and n is an integer. For example,

$$\log_{10}15 = \log_{10}(10 \times 1.5) = \log_{10}10 + \log_{10}1.5 = 1 + 0.1761$$

$$\log_{10}150 = \log_{10}(100 \times 1.5) = 2.1761$$

$\log_{10}0.15 = \log_{10}(0.1 \times 1.5) = -1 + 0.1761$. $-1 + 0.1761$ is written in the notation 1.1761.

The integer part of the logarithm is called the characteristic and the decimal fraction is the *mantissa*. Natural logarithms (Napierian logarithms) use the base e = 2.718 28 ..., $\log_e x$ is often written as Inx.

logarithmic function: The function $\log_e x$, where a is a constant. It is defined for positive values of x.

logarithmic scale: 1. A line in which the distance, x, from a reference point is proportional to the logarithm of a number. For example, one unit of length along the line might represent 10, two units 100, three units 1000, and so on. In this case the distance x along the logarithmic scale is given by the equation; $x = \log_{10}a$. Logarithmic scales form the basis of the slide rule since two numbers may be multiplied by adding lengths on a logarithmic scale ($\log(a \times b)$ $\log a + \log b$).

The graph of the curve $y = x^n$, when plotted on graph paper with logarithmic coordinate scales on both axes (log-log graph paper), is a straight fine since logy = $n$logx. This method can be used to establish the equation of a non-linear curve. Known values of x and y are plotted on log-log graph paper and the gradient n of the resulting line is measured enabling the equation to be found.

2. Any scale of measurement that varies logarithmically with the quantity measured. For instance, pH in chemistry is a measure of acidity or alkalinity – i.e. of hydrogen-ion concentration. It is defined as $\log_{10}(1/[H^*1])$. An increase in pH from 5 to 6 represents a decrease in (H^+) from 10^{-5} to 10^{-6} i.e. a factor of 10. An example of a logarithmic scale in physics is the decibel scale used for noise level.

logic The study of the methods and principles used in distinguishing corrector valid arguments and reasoning from incorrect or invalid argument, and reasoning. The main concerts in logic is not whether a conclusion is in fact accurate, but whether the process by which it is derived from a set of initial assumptions (premisses) is correct. Thus, for example, the following form of argument is valid:

all A is B

all B is C

therefore all A is C,

and thus the conclusion

all fish have wings

can be derived, correctly, from the premisses

all fish are mammals

and mammals have wings

even though site premise, and the conclusion are untrue similarly, true premisses and a true are no guarantee of a valid argument. The conclusion

all cats are mammals

does not follow, logically, from the true premisses:

all cats are warm-blooded

and

all mammals are warm blooded

because it is an example of the invalid argument form

all A is B

all C is B

therefore, all A is C.

The incorrectness of the argument shows up clearly when, after making reasonable substitutions for A, B, and C, we get true premisses but a false conclusion.

all dogs are mammals

all cats are mammals

therefore all dogs are cats.

Such an argument is called a fallacy.

Logic puts forward and examines rules that will ensure that -

given true premisses–a true conclusion can automatically be reached. It is not concerned with examining or assessing the truth of the premisses; it is concerned with the form and structure of arguments, not their content.

Logic circuit: An electronic switching circuit that performs a logical operation such as 'and' or 'limplies' on its input signals. There are two possible levels for the input and output signals, high and low, sometimes indicated by the binary digits 1 and 0, which cars be combined like the values 'true' and 'false in a truth table. For example, a circuit with two inputs and one output might have a high output only when the inputs are different. The output therefore is the logical function 'either...or...' of the two inputs (the exclusive disjunction).

logic gate: An electronic circuit that carries out logical operations. Examples of such operations are and', 'either–or 'not', 'neither nor', etc. Logic gates operate on high or low input and output voltage. Binary logic circuits, those that switch between two voltage levels (high and low), are widely used in digital computers. The *inverter* gate of *NOT* gate simply changes a high input to, a low output and vice versa. In its simplest form the *AND* gate has two inputs and one output. The output is high if and only if both inputs are high. The *NAND* gate (out and) is similar, but has the opposite effect; that is, a low output if and only if both inputs are high. The *OR* gate bas a high output if one or more of the inputs are high. The exclusive *OR* gate has a high input only if ont of the inputs, but not more than one, is high. The *NOR* gaie has a high output only if all the inputs are low. Logic gates are constructed using transistors, but in a circuit diagrant they are often shown by symbols that denote only their logical functions These fruictions are. in effect, those relationhip that can hold between propositions in symbolic logic, and whose combination can be represented in a truth table.

log-linear graph (semlogari-thmic graph) A graph on which ont axis bas logarithmic scale and the other has a linear scale. On a log-linear graph, an exponential function (one of the form $y = ke^{ax}$ where k and y are constants) is a straight line values of y are plotted on the linear scale and values of x on the logarithmic scale.

longitude: The eastwest position of a point on the Earth's surface measured as the angle in degrees from a standard meridian (taken as the Greenwich meridian). A meridian is a great circle that passes through the North and South poles. The local meridian of a pointgreat circle passing through that point and the two poles.

longitudinal wave: A wavc motion in which the vibrations in the medium are in the same direction as the direction of energy transfer. Sound waves, transmitted by alternate compression and rarefaction of the medium, are an example of longitudinal waves. Compare transverse wave.

loops: A sequence of instructions in a computer program that is performed either a specified number of times or is performed reptatedly until or some condition is satisfied.

Lorentz -Fitzgerald contraction: A reduction in the length of a body moving with a speed v relative to an observer, as compared with the length of an identical object al test relative to the observer. The object is supposed to contract by a factor Ö/(1 – v^2/ c^2), c being the speed of light in free space. The contraction was postulated to account for the negative result of the Michelson-Morley experiment using the ideas of classical physics. The idea was made it was that the electromagnetic forces holding atoms together were modified by motion through the ether. The idea was made superfluous (along with the concept of the ether) by the theory of relativity, which supplied an alternative explanation of the Michelson Morley experiment.

lumen: Symbol; lm The SI unit of luminous flux, equal to the luminous flux emitted by a point source of one candela in a solid angle of one steradian. 1 cm = 1 cd sr.

lune A portion of the area of a sphere bounded by two great semicircles that have common end points.

lux Symbol: lx The SI unit of illumination, equal to the illumination produced by a luminous flux of one lumen falling on a surface of one square metre. 1 lx = 1 m m^{-2}.

M

machine: A device for transmitting force or energy between one place and another. The user applies a force to the machine; the machine applies a force to something. These two forces need not be the same; in fact the propose of the machine to overcame a large load with a small effort. For any machine this relationship is measured by the force ratio the force applied by the machine divided by the force applied by the user (effort, F_1).

The work done by the machine cannot exceed the work done to the machine. Therefore, for a 100% efficient machine:

$$\text{if } F_2 > F_1 \text{ then } s_2 < s_1$$

$$\text{and if } F_2 < F_1 \text{ then } s_2 > s_1$$

Here s_2 and s_1 are the distances moved by F_2 and F_1 in a given time.

The relationship between si and s_2 in a given case is measured by the distance ratio (or velocity ratio) the distance moved by the effort (s_1) divided by the distance moved by the load (s_2).

Neither distance ratio nor force ratio have a unit; neither has a standard symbol.

magnetic tape: A long strip of flexible plastic with a magnetic coating on which information can be stored. Its use in the recording and reproduction of sound is well-known. It is also widely used in computing to store information that is to be fed into a computer and that is obtained from a computer during and after a program has been run. The data is stored on the tape in the forms of small closely packed magnetic spots arranged in rows across the tape. The spots are, magnetized in one of two directions that the data is in binary form.

The magnetization pattern of a row of spots represents a letter, digit (0-9), or some other character. There are usually nine or seven spot positions across a tape, forming columns or *tracks* along its length. A number of adjacent rows is used to store an item of information, With 800, 1600, or 6250 rows pet inch of tapes one magnetic tape can store an immense amount of information, This information can be altered or deleted as necessary by magnetic means. A tape can therefore be reused many times. The tape must be of very high quality and is typically 1/2 inch (1.27 cm) wide and can lie up to 700 metres long. It is generally wound on reels; cassettes are also used.

Information can be recorded on tape using a special typewriter; this method is know as key-to-tape The information is fed into a computer using a magnetic tape unit (MTU), In the simplest version two tape reels are driven at very high speed so that a magnetic tape is wound from one reel to the other and hack again. As it does so each track on the tape passes close to a small electromagnet, called a *read-write* head. These extract (read) information required by the central processor of the computer, according to Program instructions, and record (write) information that is sent from the central processor. A particular Piece of information can only be read or written when the tape has been wound into position under the heads. The MTU is therefore a serial-access as opposed to a random-access device. It is widely used as a backing store.

magnitude: 1. The absolute value of a number (without regard to sign).

2. The non-directional part of a vector, corresponding to the length of the line representing it.

Markov chain: A sequence of discrete random events or variables rital have probabilities depending on previous event in the chain.

mass Symbol: m A measure of the quantity of matter in an object. The SI unit of mass L the kilogram. Mass is determined in two ways: the inertial mass of a body determines its tendency to resist change in motion; the gravitational mm determines its gravitational attraction for other masses.

mass-energy equation: The equation $E = mc^2$, where E is the total energy (test mass energy + kinetic energy + potential energy) of a mass m, c being the speed of light in free space. The equation is a consequence of Einstein's special theory of relativity. The equation is a quantitative expression of the idea that mass is a form of energy and energy also has mass. Conversion of rest-mass energy into kinetic energy is the source of power in radioactive substances and the basis of nuclear-power generation.

matrix: A set of quantities arranged in rows and columns to form a rectangular array. The common notation is to enclose these in parentheses. Matrices do not have a numerical value, like determinants. They are used to represent relations between the quantifies. For example, a plane vector can be represented by a single column matrix with two numbers, a 2 × 1 matrix, in which the upper number represents its component parallel to the x-axis and the lower number represents the component parallel to the y-axis. Matrices can also be used to represent, and solve, simultaneous equations. In general, an $m \times n$ matrix–i.e. one with in rows and n columns is written with the first row:

$$A = a_{11}\ a_{12} \dots a_{1n}$$

The second row is:

$$a_{21}\ a_{22} \dots a_{2n}$$

and so on, the mth row being:

$$a_{m1}\ a_{m2} \dots a_{mn}$$

The individual quantities a_{11} a_{21} etc., are called elements of the matrix. The number of rows and columns, m x n, is the order or dimensions of the matrix. Two matrices are equal only if they are of the same order and if all their corresponding elements are equal. Matrices, like numbers, can be added, subtracted, multiplied, and treated algebraically accord-ing to certain rules. However, the commutative, associative, and distributive laws of ordinary arithmetic do trot apply. Matrix addition consists of adding corresponding elements together to obtain another matrix of the same order. Only matrices of the same order can be added. Similarly, the result of subtracting two matrices is the matrix formed by the differences between corres-ponding elements. Matrix multiplication also has certain rules. In multiplica-tion of an $m \times n$ matrix by a number or constant k, die result is another $m \times n$ matrix, if the element in the *ith* row and *jth* column is aij then the corresponding element in the product is ka_{ij}. This operation is distributive river matrix addition and subtraction, that is, for two matrices A and B,

$$A(A + B) = kA + kB$$

Also, $kA = Ak$, as for multiplication of numbers. In the multiplication of two matrices, the matrices A and B can only be multiplied together to form the product AB if the n umber of columns in A is the same as the number of rows in B. In ibis case they are called conformable matrices. If A is an $m \times p$ matrix with elements a_{ij} and B is a $p \times n$ matrix with elements b_{ij} then their product $AB = C$ is a n $m \times n$ matrix with elements (C_{ij}, such that C_{ij} is the sum of the products

$$a_{ij}b_{ij}+a_{i2}\,b_{2j}+a_{i3}3b-3_{j}+a_{ip}b_{pj}$$

Matrix multiplication is not commutative, that is, AB + BA.

maximum point: A point on the graph of a function at which it has the highest value within an interval. If the function is a smooth continuous curve, the maxi mum is a turning point, that is, the slope of the tangent to the curve changes continuously from positive to negative by passing through zero. If there is a higher value of the function outside the immediate neighbourhood of the maximum, it is a local maximum. If it is higher than ail other values of the function it is an absolute maximum.

mean: A representative or expected value for a set of numbers. The arithmetic mean or average (called mean) of

$$x_i, x_2, ...x_n$$

is given by:

$$(x_1, x_2, x_3, ...x_n)/n$$

In If $x_1, x_2,.....x_k$ occur with frequencies $x_1, x_2,.....x_k$ then the arithmetic mean is

$$(f_1x_1+f_2x_2+.....f_kx_k)/(f_1+f_2+...+f_k)$$

When data is classified, as in a frequency table, x_1 is replaced by the class mark.

The weighted mean W =

$$(w_1x_1+w_2x_2+....+w_nx_n)/(w_1+w_2+....+w_n)$$

where weight w_1 is associated with x_i.

The harmonic mean H =

$n/[(1/x_1) + (1/x_2) + ... (1/x_n)]$. The geometric mean $G = (x_1, x_2 .... x_n)^{1/n}$

The mean of a randorn variable

mean deviation: A measure of the dispersion of a set of numbers. It is equal to the average, for a set of numbers, of the differences between each number and the set's mean value If x is a random variable with mean value m, then the deviation is the average, or expected value, of $|x - \mu|$ written:

$$|x_i - m/n$$

mechanics: The study of forces and their effect on objects. If the forces on an object or in a system cause no change of momentum, the object or system is in equilibrium. The study of such cases is statics, If the forces acting do change moment= the study is of *dynamics*. The ideas of dynamics relate the forces to the momentum changes produced. Kinematics is the study of motion without consideration of its cause.

median: The middle number of a set of numbers arranged in order. When there is an even number of numbers the median is the average of the middle two. For example, the median of 1,3,5,11,11 is 5, and of 1,3,5,11,11,14 is (5 + ll)/2 = 8. The median of a large population is the 50th percentile (P_{50}).

2. In geometry, a straight line joining the vertex of a triangle to the midpoint of the opposite side. The medians of a triangle intersect at a single point, which is the centroid of the triangle.

mega: Symbol: M A prefix denoting 10^6. For example, 1 megahertz (MHz) 10^6 hertz (Hz).

mensuration: The study of measurements, especially of the dimensions of geometric figures in order to calculate their areas and volumes.

Mercator's Projection: A method of mapping points front the surface of a sphere onto a plane surface. Mercator's Projection is used. In make maps of the World. The lines of longitude on the sphere become straight vertical fines on the plane. The lines of latitude on the sphere become straight horizontal lines. Areas further from the equator are more stretched out in the horizontal direction. For a particular point on the surface of the sphere ai angle q latitude and angle f longitude, the corresponding Cartesian coordinates on the map are:

$$x = k\theta$$
$$y = k \log \tan(\phi/2).$$

Mercator's projection is an example of a conformal mapping, in which the angles between lines are preserved.

Metre Symbol: m The SI base unit of length, defined as the length equal in 1650763.73 wavelengths in vacuum corresponding to the transition between the levels $2p_{10}$ and $5d_5$ of the krypton-86 atom.

metric space: Any set of points, such as a plane or a volume in geometrical space, in which a pair of Points a and b with a distance d(a,b) between them, satisfy the conditions that $d(a,b) \geq 0$ and d(a,b) 0 if and only if a and b are the same point. Another property of a metric space is that $d(a,b) + d(b,c) \rightarrow d(a,c)$. The set of all the functions of x that are continuous in the interval $x = a$ to $x = b$ is also a metric space.If f(x) and g(x) are in the space,

$$[f(x) - g(x)]dx$$

is defined for ail values of x between a and b, and the integral is zero if and only if f(x) g(x) for all values of x between a and b.

metric system: A system of units based on the metre and the kilogram and using multiples and submultiples of 10. SI units, e.g.s. units, and m.k.s. units are ail scientific metric systems of units

metrology: The study of units of measurements and the methods of making precise measurements. Every physi-cal quantity that can be quantified is expressed by a relationship of the type $q = nu$, where q is the physical quantity, a is a number, and u is a unit of measurement. One of the Prime concerns of metrologists is to select and define units for ait physical quantities.

Michelson-Morley experiment: A famous experiment to detect the ether, the medium that was supposed to be necessary for the transmission of electro-magnetic waves in free space.

In the experiment, two light beams were combined to produce interference pat-terns after travelling for short equal distances perpendicular to each other. The apparatus was then turned through 90, and the two interference patterns were compared to sec if there had been a shift of the fringes. If light has a velocity relative to the ether and there is an ether 'wind' as the Earth moves through space, then the titres of travel of the two, beams would change, resulting in a fringe shift. No shift was detected even when the experiment was repeated six months later when the ether wind would have reversed direction.

mil: 1. A unit of length equal to one thousandth of an inch, It is commonly called a 'thou' and is equivalent to 2.54×10^{-5} m.

2. A unit of area, usually called a circular *mil*, equal to the area of a circle having a diameter of 1 mil.

minimum point: A point on the graph a function at which it has the lowest value within an interval. If the function is a smooth continuous curve, the minimum is a turning point, that is, the slope of the tangent to the curve changes continuously from positive to negative by passing through zero. If there is a lower value of the function outside the immediate neighbourhood of the minimum. If it is lower than all other values of the function it is an absolute minimum.

minuend: The term, from which another term is subtracted in a difference. In 5 – 4 = 1, 5 is the minuend (4 is the subtrahend).

minute: (of arc) A unit of plane angle equal to one sixtieth of a degree.

m.k.s. system: A system of units based on the metre, the kilogram, and the second. It formed the basis for SI units.

mmHg: (millimetre of mercury) A former unit of pressure defined as the pressure that will support a column of mercury one millimetre high under specified conditions. It is equal to 133.3224 Pa. It is equivalent to the torr.

Mode: The number that occurs most frequently in a set of numbers. For example, the mode (modal value) of (5, 6, 2 3, 2, 1, 2) is 2. If a continuous random variable has probability density function f(x), the mode is the value of x for which f(x) is a maximum. If such a variable has a frequency curve that is approximately symmetrical and has only one mode, then (mean mode) 3 (mean median).

Modulus: The absolute value of a quantity, not considering its sign or direction. For example, the modulus of minus five, written $|-5|$, is 5. The modulus of a vector quantity corresponds to the length Or magnitude of the vector. The modulus of a complex number x + iy is $\sqrt{(x^2+y^2)}$. If the number is written in the form $r(\cos\theta + i\sin\theta)$, the modulus is r.

mole: Symbol mol The SI base unit of amount of substance, defined as the amount of substance that contains as many elementary entities as there are atoms in 0.012 kilogram of carbon-12. The elementary entities may be atoms, molecules, ions, electrons, photons, etc., and they must be specified. One mole contains $6.022\ 52 \times 10^{23}$ entities. One mole of an element with relative atomic mass A has a mass of A grains (this was formerly called one gram-atom). One mole of a compound with relative molecular mass M has a mass of M grams (this was formerly called one gram-molecule).

moment: The turning effect produced by a force about a point, fi the Point lies on the line of action of the force the moment of the force is zero. Otherwise it is the product of the force and the perpendicular distance front the point to the fine of action of the force. If a number of forces are acting on a body, the resultant moment is the algebraic sum of ail the individual moments. For a body in equilibrium, the sum of the clockwise moments is equal to the sum of the anticlockwise moments (this law is sometimes called the law of moments)

moment generating function: (m.g.f.) A function that is used to calculate the statistical properties of a random variable x. It is defined in terms of a second variable, t, such that the m.g.f., M(t) is the expectation value of e^{tx},E (e^{tx}). For a discrete random variable

$$M(t) = -\Sigma e^{tx} p$$

and for a continuous random variable

$$M(t) = \int e^{tx} f(x).dx$$

The distributions are the same if their m.g.f.s are the same. The mean and variance of a distribution can be found by differentiating the m.g.f. The mean $E(x)\ M'(O)$ and the variance, $\text{Var}(x) = M'(O)\ (M''(O))^2$

moment of area: For a given surface, the moment of area, is the moment of mass that the surface would have if it had unit mass per unit area.

moment of inertia: Symbol I The rotational analogue of mass. The moment of inertia of an object rotating about an axis is given by

$$I = mr^2$$

where m is the mass of an element a distant r from the axis.

moment of mass: The moment of mass of a point mass about a point, fine, or plane is the product of the mass and the distance front the point or the mass and the perpendicular distance from, the line or plane. For a system of point masses, the moment of masses is the sum of the mass-distance products for the individual masses. For an object the integral must be used over the volume of the object.

momentum, linear: Symbol p The product of an object's mass and its velocity: p = mv. The object's momentum cannot change unless a net outside force acts. This relates to Newton's laws and to the definition of force. It also relates to the principle of constant momentum.

monotonic: Always changing in the same direction. A monotonic *increasing* function of a variable x increases or stays constant as x increases, but never decreases. A monotonic decreasing function of x decreases or stays constant as x increases, but never increases. Each term in a monotonic series is, in an increasing monotonic, grea-ter than or equal to the one before it; or, in a decreasing monotonic less than or equal to the one before it.

multiple: A number of expression that has a given number or expression as a factor.

multiple point: A point on the curve of a function at which several arcs inter sect, or which forms an isolated point, and where a simple derivative of the function does not exist. If the equation of the curve is written in the form:

$$(a_1x + b_1y) + (a_2x^2 + b_2xy + c^2y^2 + (a^3x^3 + \ldots = 0$$

in which the multiple point is at the origin of a Cartesian-coordinate system, the values of the coefficients of x and y indicate the type of multiple point. If a_1 and b_1 are zero, that is, if ail the first degree terms are zero, then the origin is a singular point. If the terms a_2, b_2, and c_2 are also zero, it is a double point. If, in addition, the coefficients a_3, b_3 etc., of the third degree terms; are zero, it is a triple point, and so on.

multiplicand: A number or term that is multiplied by another (the multiplier) in a multiplication.

multiplication Symbol: × The operation of finding the product of, two or more quantities, In arithmetic, multiplication of one number, a, by another, b, consists of adding a to itself b times. This kind of multiplication is commutative, that is, $a \times b = b \times a$. The identity element for arithmetic multiplication is 1, i.e., multiplication by 1 produces no change. In a series of multiplications, the order in which they are carried out does not change the result.

N

Napier formula: A set of equations used in spherical trigonometry to calculate the sides and angles in a spherical triangle. In a spherical triangle with sides *a*, *b*, and *c*, and angles opposite these of α, β and γ respectively:

$$\sin \tfrac{1}{2}(a-b)/\sin\tfrac{1}{2}(a+b) =$$

$$\tan \tfrac{1}{2}(\alpha,- \beta)/\tan \tfrac{1}{2}\gamma$$

$$\cos \tfrac{1}{2}(a-b) \cos \tfrac{1}{2}(a+b) =$$

$$\tan \tfrac{1}{2} (\alpha,+ \beta) \tan \tfrac{1}{2}\gamma$$

$$\sin \tfrac{1}{2}(\alpha,- \beta)/\sin \tfrac{1}{2}(\alpha,+\beta) = \cos \tfrac{1}{2}c$$

$$\tan \tfrac{1}{2} (a-b)/\cot \tfrac{1}{2} c$$

$$\cos \tfrac{1}{2}(\alpha,- \beta)/\cos \tfrac{1}{2}(\alpha,+\beta) \cot \tfrac{1}{2}c$$

$$\tan \tfrac{1}{2}(a-b)/\cot \tfrac{1}{2}c$$

nappe: One of the two parts of a cortical surface that lie either side of the vertex.

natural frequency: The frequency at which an object or system will vibrate freely. A free vibration occurs when there is no external periodic force and little resistance. The amplitude of free vibrations must bee too great, For instance, a pendulum swinging with small swings under its own weight moves at its natural frequency. Normally, an object's natural frequency is its fundamental frequency.

natural numbers: Symbol N The set of numbers {1, 2, 3,...} used for counting separate objects.

negation Symbol: or In logic, the operation of Putting no(or if is not the case that in front of a proposition or statement, thus reversing its truth value. The negation of a Proposition p is false if P is true and vice versa. The truth table definition for negation.

negative: Denoting a number of quantity that is less than zero, Negative numbers are also used to denote quantities that are below sortie specified reference point. For example, in the centrigrade temperature scale a temperature of –24°C is 24°C below the freezing point of water.

nesting: The embedding of a computer subroutine or a loop of instructions within another subroutine or loop, which in turn may lie within yet another, and so on.

net: 1. Denoting a weight of goods excluding the weight of the containers or Packing.

2. Denoting a profit calculated after deducting all overhead costs, expenses, and taxes. Compare gross.

neutral equilibrium: Equili-brium such that if the system, is disturbed a little, there is no tendency for it to move further nor to return.

newton Symbol: N The SI unit of force, equal to the force needed to, accelerate one kilogram by one meter second-2. 1 N = 1 kg m s-2.

Newtonian mechanics: Mechanics based on Newton's laws of motion; i.e. relativistic effects are not taken into account.

Newton's law of universal gravitation: The force of gravitational attraction bet-ween two point masses (m_1 and m_2) is proportional to, each mass and inversely proportional to, the square of the distance (r) between them. The law is often given in the form

$$F = Gm_1m_2/r^2$$

where G is a constant of proportionality called the gravitational constant. The law can also be applied to, bodies; for example, spherical objects can be assumed to have their mass acting at the centre.

Newton's laws of motion: Three laws of mechanics formulated by Sir Isaac Newton in 1687. They can be stated as: (1) An object continues in a state of test or constant velocity unless acted on by an external force. (2) The resultant force acting on an object is proportional to the rate of change of momentum of the object, the change of momentum being in the same direction as the form (3) If one object exerts a force on another then there is an equal and opposite force (reaction) on the first object exerted by the second.

Newton's method: A technique for obtaining successive approximations. (iterations) to the solution of air equation, each more accurate than the preceding one. The equation in a variable x is written in the form f(x) = 0, and the general formula or algorithm:

$$x_{n+1} = x_n - f(x_n)/f'(x_n)$$

is applied, where xn is the nth approximation. New-ton's method can be thought of as repeated estimates of the position on a graph of f(x) against x at which the curve crosses the x-axis by extrapolation of the tangent to, the curve. The slope of the tangent at (x_1, f) (x_1) is df/dx at $x - x_1$ that is

$$f'(x_1) = f(x_1)/(x_2 - x_1)$$

$x_2 = x_1 - f(x_1)/f(x_1)$ is therefore the point where the tangent crosses the x-axis, and is a closer approximation to x at $f(x) = 0$ than x_1 is. Similarly,

$$x_3 = x_2 - f(x_2)/f'(x_2)$$

for example, if $f(x) = x^2 - 3 = 0$, then $f(x) = 2x$ and we obtain the algorithm $x_{n+1} = x_n - (x_n^2 - 3)/2$ $(x_n = ½\ (x_n + 3/x_n)$.

node: A point of minimum vibration in a stationary wave pattern, as near the closed end of a resonating pipe. Compare antinode.

nominal value: (per value) The value given to a stock or share by the government or limited company that offers if for sale. Stocks invariably have a nominal value of £100. Shares, however, may have any nominal value. For example, a company wishing to raise £100 000 by an issue of shares may issue 100 000 £1 shares or 200 000 50p shares, or any other combination. The issue price, i.e. the price paid by the first buyers of the shares, may not be the same as the nominal value, although if is likely to be close to it. A share with a nominal value of 50P may be offered ai an issue price of 55p; it is then said to be offered ai a premium of 5p. If offered ai an issue price of 45p it is said to be offered at a discount of 5p. Once established as a marketable share on a stock exchange, the nominal value has little importance and it is the market price ai which it is bought and sold. However, the dividend is always expressed as a Percentage of the nominal value.

numogram: A graph that consists of three parallel lines, each one a scale for one of three related variables. A straight line drawn between two Points, representing known values of two of the variables, crosses the third line ai the corresponding value of the third variable. For example, the lines might show the temperature volume, and pressure of a known mass of gas. If the volume and pressure are known, the temperature can be read off the nomograrn.

nonagon: A plane figure with nine straight sides. A regular has nine equal sides and nine angles.

non-Euclidean geometry: Any system of geometry postulate of Euclid does not old. This postulate states that if a point lies outside a straight line only one line parallel to the straight line can be drawn through the point. In the early nineteenth century it was shown that it is possible to have a whole self-consistent formal system of geometry without using the parallel postulate at all. There are two types of non-Euclidean geometry. In one (called elliptic geometry) there are no parallel fines through the point. An example of ,this is a logical system describing the properties of lines, figure, angle~, etc., on the ~surface of a sphere, in which all lines are parts of great circles (i.e. circles of the sphere). Since all great circles intersect, no parallel can be drawn through the point. Note also that the angles of a triangle on such a sphere do not add up to 180°. The other type of non-euclidean geometry is called a hyperbolic geometry here an infinite number or paralles can be drawn through the point.

Note that a type of geometry is not itself based on 'experiment' i.e., measure-ments of distance, angles, etc. It is a purely abstract system based on certain assumptions. Mathema-ticians study such systems for their own sake without necessarily looking for practical applications. The practical applications come in, when a particular mathematical system gives an accurate description of physical properties i.e., the properties of the 'real world'. In practical uses it is assumed that Euclidean geometry applies. However it is found that ibis is only an approximation, and that the space-time continuum of relativity theory is non-Euclidean in its properties.

normal: Denoting a fine or plane that is perpendicular to another line or plane, A line or plane is said to be normal to a curve if it is perpendicular to the tangent to the curve, at the point ai which the line and the curve meet A radius of a circle, for example, is normal to the circumference. A plane passing through the centre of a sphere is normal to the surface ai ail points at which they meet.

normal distribution: (Gaussian distribution) The type of statistical distribution followed by, for example, the same measurement taken several times, where the variation of a quantity (x) about its mean value (μ) is entirely random. A normal distribution has the probability density function.

$$f(x) = \exp\ [-(x - \mu)^2\ 2\sigma^2\ ]/\sigma 2\pi$$

where α is known as the standard deviation. The distribution is written N ($\mu\sigma^2$). The graph of $f(x)$ is bell-shaped and symmetrical about $x = \mu = 0$. The standard normal distribution has and $\sigma^2 = 1.x$ can be standardized by letting $z = (x - \mu)/\sigma$. The values for which the area under the curve from to z_a is are tabulated.

null matrix: (zero matrix) A matrix in which ail the elements are equal to zero.

number line: A straight horizontal line on which each point represents a real number, Integers are points marked ai unit distance apart.

numbers: Symbols used for counting and measuring. The numbers now in general use are based on the Hindu-Arabic system, which was introduced to Europe in the 14th and 15th centuries. The Roman numerals used before this made simple arithmetic very difficult, and most calculations needed an abacus. Hindu-Arabic nume-rals (0, 1, 2, ... 9) enabled calculations to be performed with far greater efficiency because they are grouped systematically in units, tens, hundreds, and so on.

numerator: The top part of a fraction. For example, in the fraction 3/4, 3 is the numerator and 4 is the denominator.

numerical analysis: The study of methods of calculation that involve approximations, for example, iterative methods.

numerical integration: A procedure for Calculating approximate values of integrals. Sometimes a function is known only as a set or values of a corresponding Values of a variable and not as a general formula that can be integrated. Usually, a calculation is first carried out with a few columns; these are further subdivided until the desired accuracy is attained, i.e., when further subdivision makes no significant difference to the result.

O

object: The set of points that undergoes a geometrical transformation or mapping.

oblate: Denoting a spheroid that has a polar diameter that is smaller than the equatorial diameter. The Earth, for example, is not a perfect sphere but is oblate spheroid.

oblique: Forming an angle that is not a angle.

oblique solid: A solid geometrical figure that is 'slanted'; for example, a cone, slander, pyramid, or prism with an axis at is not at right angles to its base compare right solid

oblique triangle: A triangle that dues not obtain a right angle.

obtuse: Denoting an angle that is greater than a right angle; i.e. greater than 90° (or $\pi/2$ radian).

OCR: (optical character recognition) A system used to input information to computer. The information, usually into form of letters and numbers, is printed, typed, or sometimes handwitten. The characters used can be ready people and can also be read and identified optically by an OCR reader. This machine interprets each character and translates it into a series of electrical pulses. The pulses can then be transmitted to the central processor of the computer.

octagon: A plane figure with eight straight side. A regular octagon has eight equal sides and eight equal angles.

octahedron: A polyhedron that has eight faces. A regular octagon has eight faces, each one an equilateral triangle.

octal: Denoting or based on the number eight An octal number system but eight different digits instead of the ten in the decimal system. Eight is written as 10 nine as 11, and so on.

octant: 1. One of eight regions into which space is by the three axes of a three-dimensional Cartesian coordinate system. The first octant is the on in which x, y, and z are all positive. The second, third, and fourth octants are numbered anticlockwise around the positive z-axis. The fifth octant is underneath the first, the sixth under the second, etc. 2. A unit of plane angle equal to 45 degrees ($\pi/4$ radians).

odd: Not divisible by two. The set of odd numbers is (1, 3, 5, 7, ...).

odd function: A function f(x) of a variable x for which f($-x$) f($-x$), For example, nx and x_1 are odd functions of x^3.

oersted Symbol: Oe A unit of magneticfield strength in the c.g.s. system. It is equal to $10^3/4\pi$ amperes per metre ($10^3/4\pi$ A m^{-1}).

off-line: Disconnected front or net trader the direct control of a computer. Off-line equipment is either not in use, is undergoing repair, or is performing some task without the assistance of the central processor of the computer.

ohm: Symbol Ω. The SI unit of electrical resistance, equal to, a resistance that passes a current of one ampere when there is an electric potential difference of one volt across it. 1Ω = 1ς A^{-1}. Formerly, it was defined in terms of the resistance of a column of mercury under specified conditions.

one-to-one, correspondence: A function or mapping bet-ween two sets of things or numbers, such that each element in the first set maps into only one element in the second, and vice versa. For example, the set of left, ears and the set of right ears are in a one-to- one corres-pondence. The set of fathers and the set of sons are not.

on-line: Connected to and controlled by a computer. Any device that is connected to and, capable of interacting directly with a computer without human intervention is said to be on-line. In on-line Processing the processing of a computer program is performed on equipment directly control-led by the central processor.

open curve: A curve in which the ends do not meet, for example, a parabola or a hyperbola.

open interval: A set consisting of the numbers between two given numbers (end points), not including the end points, for example, all the real number greater than 1 and less than 4.5. The open interval between two real numbers a and b is written (a, b). On a number line, the end points of an open interval are circled.

open set: A set defined by limits that are not included in the set itself. The set of all rational numbers greater than 0 and less than ten, written (x: $0 < x < 10$; $x \varepsilon R$), and the set of all the points inside a circle, but not including the circle itself, art examples of open sets.

operating system (OS): The collection of programs used in the control of a computer system. It is generally supplied by the computer manufacturer. An operating system has to decide at any instant which of the many demands, on the attention of the Central processor to satisfy next. These demands include input from and output to various the execution of a number of programs and accounting and timing. Large computers, in which many job can run simultaneously, will have a highly complex operating system; small computers may have a very simple program that runs without the benefit of an operating system is called a alone program.

operator: 1. A mathematical function such as addition subtraction, multiplication, or taking a square root or a logarithm, etc. 2 The symbol denoting a mathematical operation or function, for example: +, –. ×, √, $\log_{10}$.

opposite: Denoting the side facing a given angle in a triangle, i.e. the side not forming one arm of the angle. In trigonometry, the ratios of the length of the opposite side to the other side lengths in a right-angled triangle are used to define the sine and tangent functions of the angle.

orbit: The curved path or trajectory along which a moving object travel under the influence of a gravitational field. An object with a negligible mass moving under the influence of a planer or other body has an orbit that is a conic section: i.e., a parabola, ellipse, or hyperbola. In considering the motion of planets, a correction must be made for the planet's mass.

order: 1. (of a matrix) The number of rows and columns in a matrix. 2. (of a derivative) The number of times a variable is differentiated. For example, dy/dx is a first-order derivative; d^2y/dx^2 is second order; etc. 3. (of a differential equation). The order of the highest derivative in an equation. For example,

$$d^3y/dx^3 + 4xd^2y/dx^2 = 0$$

is a third order differential equation.

$$d^1y/dx^2 - 3x(dy/dx)^3 = 0$$

is a second-order differential equation.

ordered pair: Two numbers indicating values of two variables in a particular order. For example, the x- and y-coordinates of points in two-dimensional Car-tesian coordinates form a set of ordered pairs (x, y).

ordered triple: Three numbers indicating values of three variables in a particular order. The x, y, and z coordinates of a point in a three-dimensional coordinate system form an ordered triple (x, y, z).

ordinal numbers: Whole numbers that denote order, as distinct from number or quantity. That is, first, second, third, and so on.

ordinary differential equation: An equation that contains total derivatives, but no partial derivatives.

ordinate: The vertical or y coordinate in a two-dimensional rectangular Cartesian coordinate system.

origin: The fixed reference point in a coordinate system, at which the values of all the coordinates are zero and at which the axes meet.

orthocentre: A point in a triangle that is the point of intersection of lines from each vertex perpendicular to the opposites sides. The triangle formed by joining the feet of these vertices is the triangle.

orthogonal projection: A geometrical transformation that produces an image on a fine or plane by perpendicular fines crossing the plane. If a line of length l is projected orthogonally from a plane, its angle θ to the image plane, its image length is $l\cos\theta$ image of a circle an ellipse.

oscillation: A regularly repeated motion or change.

ounce: 1. A unit of mass equal to onesixteenth of a pound. It is equivalent to 0.028349 kg. 2. A unit of capacity, often called a fluid ounce, equal to one twentieth of a pint. It is equivalent in 2.8413×10^{-5} m^3. In the USA a fluid ounce is equal to onesixteenth of a US pint. It is equivalent to 2.0573×10^{-5} m^3. 1 UK fluid ounce is equal to 0.9608 US fluid ounce.

output: 1. The signal or other form of information obtained from an electrical device, machine, etc. The output of a computer is the information or results derived from the data and programmed instructions fed into it. This information is transferred as a series of electrical pulses from the central processor of the computer to a selected output device. Some of these output units convert the pulses into a readable or pictorial form; examples include the line printer, plotter, and visual display unit (which can also be used as an input device). Other output devices translate the pulses into a form that can be fed back into the computer at a later stage; the paper-tape punch is an example. 2. The process or means by which output is obtained. 3. To deliver as output.

overdrafts: A negative balance on a current account at a bank. Interest is calculated on a daily basis. An overdraft is distinguished from a bank loan, which is for a fixed sum on all of which interest is payable either monthly or quarterly.

overflow: The situation arising in computing when a number, such as there result of an arithmetical operation, has a greater magnitude than can be represented in the space allocated to it in a register or a location in store.

overlay: A technique used in computing when the total storage requirements for a lengthy program exceeds the space available in the main store. The program is split into sections so that only the section or sections required at any one time will be transferred into the main store from a disk unit or other backing store. These program segments will all occupy the same area of the main store. The overlay structure must be organized so that no routine calls another that would overwrite it.

P

paper tape: A long strip of paper, or sometimes thin flexible plastic, on which information can be recorded as a pattern of round holes punched in rows across the tape. There arc two standard widths: 0.6875 inch and 1 inch (17.46 and 25.4 mm). The positions ai which holes may be punched are called tracks; one-inch tape with eight tracks per row is widely used. There is also a fine of small sprocket holes along the length of the tape between tracks three and four.

Paper tape is used to, input and output information in a wide variety of devices. Laboratory equipment, for example, will often produce results punched on tape. The punched information is fed into a computer using a paper-tape reader. This machine senses the presence or absence of holes in each row and converts the information into a series of electric pulses. (A hole usually produces a pulse, a non-hole produces no pulse.) The pulses are then transmitted to the central processor of the computer. Although maybe 1000 rows can be read per second, the paper-tape reader is a slow input device. Information is output on paper tape using a paper-tape punch, which automatically punches data into the tape. Punched tape from a computer can be fed back in ai a later date or fed into another computer. Compare card, magnetic tape, disk.

Pappus' theorems: Two theo-rems concerning the rotation of a curve or a plane shape about a line that lies in the same plane The first theorem states that the surface area generated by a curve, revolving about a line that does not cross it is equal to the length of the curve times the circumference of the circle traced out by its. The second theorem states that the volume of a solid of revolution generated by a plane area that rotates about a fine not crossing it, is equal to the area times the circumference of the circle traced out by the centroid of the area.

parabola: A conic with an eccentricity of 1. The curve is symmetrical about an axis through the focus at right angles to the directrix. This axis intercepts the parabola at the A chord through the focus perpendicular to the axis is the latus rectum of the parabola.

In Cartesian coordinates a parabola can be represented by an equation;

$$y^2 = 4ax$$

In this form the vertex is at the origin and the x-axis is the axis of symmetry. The focus is at the point (o,a) and the directrix is at the point (o,a) and the directrix is the line $x = -a$ (parallel to the y-axis). The latus rectum is 4a. If a point is taken on a parabola and two lines drawn from it–one parallel to the axis and the other from the point to the locus then these lines make equal angles with the tangent at that axis is known as the reflection property of the parabola, and is utilized in parabolic reflectors and antennae. The parabola is the curve traced out by a projectile fading freely under gravity. For example, a tennis ball projected horizontally a velocity v has, after time t travelled a distance $d = vt$ horizontally, and has also fallen vertically by $h = gt^2/2$ because of the acceleration A free fail g. These two equations are parametric equations of a parabola. Their standard form, corresponding to $y^2 = 4ax$, is x

$$x = at^2$$

$$y = 2at$$

where x represents h, the constant a is $g/2$, and y represents d.

Paraboloid: A curved surface in which the cross-sections in any plane passing through a central axis is a parabola. A paraboloid of revolution is formed when a parabola is rotated around its axis of symmetry. Parabolic surfaces are used in telescope mirrors, searchlights, radiant heaters, and radio antennae on account of the focusing property of the parabola.

Another type of Paraboloid is the hyperbolic paraboloid. This is a surface with the equation:

$$x^2/a^2 - y^2 = 2cz$$

where c is a positive constant. Sections parallel to the xy plane (z = 0) are hyperbolas. Sections parallel to the other two planes ($x = 0$ or $y = 0$) are parabolas.

Paradox (antinomy): A pro-position or statement that leads to a contradiction if it is asserted and if it is denied. An example is

Russell's paradox of set theory. Some sets are members of themselves (the set of sets is itself a set), some are not (the set of homes is not a horse). Consider the set that is, the set of all things that are not members of themselves. Is that set a member of itself? If it is, it is not, and if it is not, it is.

parallel: Extending in the same direction and remaining the same distance apart.

parallel forces: When the forces on an object pass through one point their resultant can be found by using the parallelogram of vectors. If the force are parallel the resultant is found by addition, taking sign into account. There, may also be a turning effect in such cases, which can be found by the principle of moments.

parallelogram: A plane figure with four straight, sides, and with opposite sides parallel and of equal length. The opposite angles of a parallelogram are also equal. Its area is the product of the length of one side and the perpendicular distance from that side to the side opposite. In the special case in which the angles are all fight angles the parallelogram is a rectangle, and when all the sides are equal it is a rhombus.

parallelogram of vectors: A method for finding the resultant of two vectors acting at a point. The two vectors are shown as two sides of a parallelogram the resultant is the diagonal through the starting point. The technique can be used either with careful scale drawing or with trigonometry. The trigono-metrical relations give:

$$F = \sqrt{(F_1^2 + F_2^2 + 2\ F_1F_2\cos\theta)}$$

$$\alpha = \sin^{-1}[(F_2/F)\sin\theta]$$

where θ is the angle between F_1 and F_2 and α is the angle between F_1 and F_2.

parallelopiped: A solid figure with six faces that are parallelograms. In a rectangular parallelopiped the faces are rectangles. If the faces are squares, the parallelopiped is a cube.

parameter: A quantity that when varied affects the value of another. For example, if a variable z is a function of variables x and y, that is z = f(x,y), then x and y are the parameters, that determine z.

parametric equations: Equa-tions that, in an implicit function (such as $f(x,y = 0)$ express x and y separately in terms of a quantity, which is an independent variable or parameter. For example, the equation of a circle can be written in the form

$$x^2 + x^2 = r^2$$

or as the parametric equations

$$x = r\cos\theta$$

$$y = r\sin\theta$$

partial derivative: The rate of change of a function of several variables as one of the variables changes and the others are held constant. For example, if $z = f(x,y)$ the partial derivative $\partial z\ \partial x$ is the rate of change of z with respect to x, with y held constant. Its value will depend on the constant value of y chosen. In three-dimensional Cartesian coordinates, $\partial z/\partial x$ is the gradient of a fine ai a tangent to the curved surface $f(x,y)$ and parallel to the x-axis.

partial differential: The infinitesimal change in a function of two or more variables resulting front, changing only one of the variables while keeping the others constant. The sum of all the partial differentials is the total differential.

partial differential equation: An equation that contains partial derivatives of a function with respect to a number of variables. General methods of solution are available only for certain types of linear partial differential equation that occur in physical problems. Laplace's equation, for example, is $\partial^2\phi\ /dx^2 + \partial\phi^2\ /\partial\ y^2 + \partial^2\ /z^2 =0$ and occurs in the study of gravitational and electromagnetic fields.

partial fractions: A sum of fractions that is equal to a particular fraction, for example, ½ + ¼ = ¾. Writing a ratio in terms of partial fractions can be useful in solving equations or calculating integrals. For example: $1/x\ (x^2+1)$ can be written as: $a/x + (bx+c)/(x^2+1)$.

The values $a = 1$, $b = 1$, and $c = 0$ are then calculated by comparing the coefficients of indentical powers of x, to give

$$1/x(x^2+1) = 1/x-1(x^2+1)$$

a form that can be integrated with, respect to x as a sum of two integrals.

partial sum: The sum of a finite number of terms from the start of an infinite series. In a convergent series, the partial sum of the first r terms, S_n is an approximation in the sum to infinity.

particle: An abstract simplification of a real object the mass is concentrated at the object's centre of mass; its volume is zero. Thus rotational aspects can be ignored.

pascal Symbol: Pa The SI unit of pressure, equal to a pressure of one newton per square meter (l Pa =1N m^{-2}). The pascal is also the unit of stress.

Pascal's distribution: (negative binomial distribution) The distribution of the number of independent Bernoulli trials performed up to and including the rth success. The probability that the number of trials, is equal to k is given by

$$P(x=k) = {}^{k-1}C^{r-1}\, pr\, q^{k-r}$$

The mean and variance are r/p and rq/p^2 respectively.

Pascal's triangle: A triangular array of numbers in which each row starts and ends with 1 and that is built up by summing two adjacent numbers in a row to, obtain the number directly below them in the next row. Each row in Pascal's triangle is a set of binomial coefficients. In the expansion of $(x+y)^n$, the coefficients of x and y are given by the $(n+1)$ th row of Pascal's triangle.

pendulum: A body that oscillates freely under the influence of gravity. A simple pendulum consists of a small mass oscillating to and fro at the end of a very light string. If the amplitude of oscillation is small (less than about 10°), it moves with simple harmonic motion; the period does not depend on amplitude. There is a continuous interchange of potential and kinetic energy through the motion; ai the ends of the swings the potential energy is maximum and the kinetic energy zero. At the mid-point the kinetic energy is a maximum and the potential energy is zero. The period is given by

$$T = 2\ \pi\sqrt{(I/g)}$$

Here *I* is the length of the pendulum (from support to, centre of the mass) and *g* is the acceleration of lice fall.

Here, k is the radius of gyration about an axis through the centre of mass and h is the distance from, the pivot to the centre of mass.

pentagon: A plane figure with five straight sides. In a regular pentagon, one with ail five sides and angles equal, the angles are all 108° A regular pentagon can be superimposed on itself after rotation through 720° (2π/5 radians).

percentage: A number expressed as a fraction of one hundred. For example, 5 percent (or 5%) is equal in, 5/100. Any fraction or decimal can be expressed as a percentage by multiplying it by 100. For example, 6.63 × 100 63% and ¼ × 100 = 25%.

percentage error: The error or uncertainty in a measurement expressed as a percentage of the total measurement. For example, if, in measuring a length of 20 meters, a tape can measure to the nearest four centimeters, the measure-ment is written as 20 ± 0.04 meters and the percentage error is (0.04/20) × 100 = 0.2%.

percentile: One cf the set of points that divide a set of data arranged in numerical order into 100 parts. The rth percentile, P is the value below and including which r% of the data lies and above which (100-r)% lies. P_r, can be found from the cumulative frequency graph.

perimeter: The distance round the edge of a plane figure. For example, the perimeter of a rectangle is twice the length plus twice the breadth. The perimeter of a circle is its circumference.

periodic motion: Any kind of regularly repeated motion, such as the swinging of a pendulum, the orbiting of a satellite, the vibration of a source of sound, or an electromagnetic wave.

peripheral unit: (peripheral) A device connected to and controlled by the central processor of a computer. Periperals include input devices, output devices, and backing store. Some examples are visual display units, fine printer, magnetic tape units, and disk units.

perpendicular: At right angles. The perpendicular bisector of a line crosses it half way along its length and forms: a right angle. A vertical surface is perpendicular to a horizontal surface.

phase: stage in a cycle that a wave (or other periodic system) has reached, at a particular time (taken from some reference point . Two waves are in phase if their maxima and minima coincide. For a simple wave represented by the equation

$$y = a\sin 2\pi(ft-x/\lambda)$$

The phase of the wave is the expression

$$2\pi(ft-x/\lambda)$$

The phase difference between two points distances x, and x: from the origin is

$$2\pi(x_1-x_2)/\lambda$$

A more general equation for a progressive wave is

$$y = a\sin\ 2\pi(ft-x/\lambda-\phi\ )$$

Here, ϕ is the phase constant the phase, when t and x are zero.

Two waves that are out of phase have different phase constants (they 'start' at different stages at the origin). The phase difference, is $\phi_1 - \phi_2$. It is equal to where x is the distance between corresponding points% on the two wave. It is the phase angle between the two waves; the angle between two rotating vectors (phasors) representing the waves.

phase speed: The speed with which the phase in a travelling wave is propagated. It is equal to where T is the period.

pi (π) The ratio of the circumferences of any circle to its diameter π is approximately equal to 3.14159... and is a transcendental number (its exact value cannot be written down, but it can be measured to any degree of accuracy.

pictogram: (pictograph) A diagram that represents statistical date in a pictorial form. For example, the number of pink, red, yellow, and white flowers that grow from a packet of mixed seeds can be shown by rows of the appropriate number of coloured flower shapes.

pie chart: A diagram in which proportions are illustrated as sectors of a circle, the relative areas of the sectors representing the different proportions. For example, if, out of 100 workers in a factory, 25 travel in work by car, 50 by bus' 10 by train, and the rest walk, the bus passengers are represented by hall of the circle, the car passengers by a quarter, the train users by a 36° sector, and so on.

pint: A unit of capacity. In the UK it is equal to one eighth of a UK gallon and is equivalent to 5,6826 $\times 10^{-4}$ m^3. The US liquid pint is equal to one eighth of a US gallon and is equivalent to 4.7318 $\times 10^{-4}$ m^3. The US dry pint is equal to one sixty-fourth of a US bushel and is equivalent to 5.5061×10^{-4} m^{-3}

plane: A flat surface, either real or imaginary, in which any two points are joined by a straight line lying entirely on the surface. Plane geometry involves the relationships between points, lines,, and curves lying in the same plane. Cartesian coordinates, any point in a plane cart be defined by two coordinates, *x* and *y*. In three-dimensional coordinates, each value of z corresponds to a plane parallel to, the plane in which the x and y axes lie. For any three points, there exists only one plane containing all three. A particular plane can also be specified by a straight line and a point.

plot: To draw on a graph. A series of individual points plotted or a graph may show a general relationship between the variables represented by the horizontal the vertical and vertical axes. For example, in scientific experiment one quantity can be represented by x and another by y. The values of y at different values of x are then plotted as a series of Points On a graph If these fall on a line or curve drawn through the points is said to be a plot of y against x.

plotter: An output device of a computer system that produces a permanent record of the results of some program by drawing lines on paper. One pen, or may be two or more pens with different coloured ink, are moved over the paper according to instructions sent front the computer or front a backing store. Plotters are used for drawing graphs, contour maps, etc.

plotting: The process of marking points on a system of coordinate or of drawing a graph by marking points.

point: A location in space, on a surface, or in a coordinate system. A point has no dimensions and is defined only by its position.

point of contact: A single point at which two, curves, or two curved surfaces meet, There is only one point of contact between the circumference of a circle and 'tangent to the circle. Two spheres also can, have only one point of contact.

point of inflection: A point or, a curved line at which the tangent changes its direction of rotation. Approaching from one side or the point of inflection the slope of the tangent to the curve increases; and moving away from it on the other side, it decreases. For example, the graph of $y = x^3 -3x^2$ in rectangular Cartesian coordinates, has a point of inflection at the point $x = 1$, $y -2$.

Poisson distribution: A proba-bility distribution for a discrete random variable. It is defined, for a variable (r) that can take values in the range 0, 1, 2, and has a mean value μ as,

$$P(r) = e^{-\mu r}/r$$

A binomial distribution with a small frequency of success p in a large number n of trials can be approximated by a Poisson distribution with mean up.

polar coordinates: A method of defining the position of a point by its distance and direction from a fixed reference point (pole). The direction from a foxed line (axis). On a flat surface only one angle, θ and the radius, r are needed to specify each point (r,θ) = (1, π/2) is the point one unit length away from the origin in the perpendicular direction. Conventionally, angles are taken as positive in the anticlockwise sense.

In a rectangular Cartesian coordinate system with the same origin and the *x*-axis at θ = 0, the *x*-and *y*- coor-dinates of the point (r,θ) are:

$$x = r\cos\theta$$

$$y = r\sin\theta$$

Conversely

$$r = \sqrt{(x^2 + y^2)}$$

and

$$\tan\theta = y/x$$

In three dimensions, two forms; of polar coordinate systems can be used.

pole 1. One of the points on the Earth's surface through its axis of rotation passes, or the corresponding point on any other sphere.

Polygon: a plane figure bounded by number of straight sides. In a regular polygon, all the sides are equal and all the internal angles are equal. In a regular polygon of n sides the exterior angle is $360°/n$.

polyhedron: A solid figure bounded by plane polygonal faces. The point at which three or more faces intersect on a polyhedron is called a vertex, and a fine along which two faces intersect is called an edge. In a regular polyhedron, ail the faces are congruent regular polygons. There are only five regular polyhedrons: the regular tetrahedron, which has four equilateral triangular faces; the regular hexahedron, or cube, which has six equilateral square faces; the regular octahedron, which has eight equilateral triangular faces; the regular dodecahedron, which has twelve regular pentagonal faces; and the regular icosahedron, which has twenty equilateral triangular faces. These are all convex polyhedrons. That is, all the angles between faces and edges are convex and the polyhedron can be laid down flat on any one of the faces. In a concave polyhedron, there is at least one face in a plane that cuts through the polyhedron, and it cannot be laid down on this face.

position vector: The vector that represents the displacement of a point front, a given reference origin. If a point P in polar coordinates has coordinates (r,θ) then r is the position vector of P a vector of magnitude *r* making an angle θ with the axis.

positive: Denoting a number or quantity that is greater than zero. Numbers that are used in counting things and measuring sizes are all positive numbers. If a change in a quantity is positive, it increases. that is it moves away from zero if it is already positive, and towards it if it is negative. Compare negative.

potential energy: Symbol V The work an object can do because of its position or state, There are many examples. The work an object at height can do in falling is its gravitational potential energy. The energy, 'stored' in elastic or spring under tension or compression is clastic potential energy. Potential difference in electricity is a similar concept, and so on. In practice the potential energy of a system is the energy involved in bringing it to its current state from' .;some reference state, or vice versa.

poundage: 1. A weight expressed in pounds. 2. A tax or charge calculated on the weight of chargeable goods in pounds or on their value in pounds sterling.

poundal: The unit of force in the system f.p.s. It is equal to 0.138 255 newton (0.138 255 N).

power: 1. The number of times a quantity is to be multiplied by itself. For example, 24 = 2 × 2 × 2 × 2 = 16 is known as the fourth *p* of two, or two to, the power four. A power series is a series of the form $a_0 + a_1x + a_2x^2 + ... + a_nx^n$.

2. Symbol P The rate of energy, transfer (or work done) by or to a system. The unit of power is the watt—the energy transfer in joules per second power series. A series in which the terms contain regularly increasing powers of a variable. For example:

$Sn = 1 + 2x + 3x^2 + 4x^3 + ... + nx^{n-1}$ is power series in the has the form $a_0 + a_,x + a^2x^2 + ... + a_nx^n$

where a_0, a_1, a_2, etc. are constants.

precession: If an object is spinning on an axis and a force is applied at right angles to this axis, then the axis of rotation can itself move around another' axis at an angle to it. The effect is seen in tops and gyroscopes, which, 'wobble' slowly while they spin as a result of the force of gravity. The Earth also precesses the axis of rotation slowly describes a cone.

The precession of Mercury is a movement of the whole orbit of the planet around an axis perpendicular to the orbital planet around an axis perpendicular to the orbital plane. It can be explained by relativistic mechanics.

precision: The number of figures in a number. For example 2.342 is stated to a precision of four significant figures, or three decimal places. The precision of a number normally reflects the accuracy of the value it represents.

premiss: In logic, an initial proposition or statement that is known or assumed to be true and on which a logical argument is based.

premium: 1. The difference between the issue price of a stock or share and its nominal value when the issue price is in excess of the nominal value.

2. The amount of money paid each year to an insurance company to purchase insurance cover for a specified risk.

pressure Symbol: *p* The pressure on a surface due to forces front another, surface or from a fluid is the force acting at 90° to, unit area of the surface: pressure = force/area.

Objects are often designed to maximize or minimize pressure applied. To give maximum pressure, a small contact area is needed as with drawing pins and knives. To give minimum pressure, a large contact area is needed as with snow shoes and the large tyres of large vehicles.

The pressure at a point at a certain depth in a fluic. (1) is the same in ail directions; (2) applies force at 90° to any contact surface; (3) does not depend on the shape of the container

pressure of the atmosphere: The pressure at a point near the Earth's surface due Io the weight of air above that point. Its value varies around about 100 kPa (100 000 newtons per square meter)

prime: Having no factors except 1. The set of prime numbers is {2, 3, 5, 7, 11, 13, 17, 19, 23, 29, ..}. The prime factors of a number are the prime numbers that divide into it exactly. For example, the prime factors of 45 are 3, 3, and 5 ($45 = 3 \times 3 \times 5$). Each whole number has a unique set of prime factors.

principal: A sum of money that is borrowed, on which interest is charged. See compound interest. simple interest.

principle of moments: The principle that when an object or system is in equilibrium the sum of the moments in any direction equals the sum of the moments in the opposite direction. Because there is no resultant turning force, the moments of the forces can be measured relative to any point in the system or outside it

printout: The computer output, in the loran of characters printed on a continuous sheet of paper, produced by a fine primer or similar device. See output, line printer.

prism: A polyhedron with two parallel opposite faces, called bases, that are congruent polygons. All the other faces, called lateral faces, are parallelograms formed by the straight parallel fines between corresponding vertices of the bases. If the bases have a centre, the fine joining the centres is the axis of the prism. If the axis is ai right angles to the base, the prism is a right prism (in which case the lateral faces are rectangles) otherwise it is an oblique pi ism. A triangular prism has triangular bases and three lateral faces. This is the shape of many of the glass prisms used in optical instruments A quadrangular prism has a quadrilateral base and four lateral faces. The cube is a special case of ibis with square bases and square lateral faces.

probability: The likelihood of a special event occurring. If an experiment has n possible and equally likely outcomes, in of which are event A, then the probability of A is P(A) m/n. For example, if A is an even number coming up when a die is thrown, then P(A) 3/6. When the probabilities of the different possible results are not already known, and event A has occurred in times in n trials, P(A) is defined as the limit of m/n as n become infinitely large.

program: A complete set of instructions to a computer, written in a programming language. (The word is also used as a verb, meaning to write such instructions) These instructions, together with the facts (usually called data) on which the instructions operate, enable the computer to perform a wide variety of tasks. For example, there are instruc-tions to do arithmetic, to move data between the main store and the central processor of the computer, to perform logical operations, and to alter the flow of control in the program.

The instructions and data must be expressed in such a way that the central processor can recognize and interpret the instructions and cause them to be carried out on the right data. They must in fact be in binary form, i.e. in a code consisting of the binary digits 0 and 1 (bits). Ibis binary code is known as machine code (or machine language). Each type of computer has its own machine code.

It is difficult, tedious, and lime consuming for people to write programs in machine code. Instead programs are usually written a source language, and these source programs are then translated into machine code, Most source programs are written in a high level language and are converted into machine code by a complicated program called a compiler. High-level languages are closer to natural language and mathematical notation than to machine code, with the instructions taking the form of statements. They are fairly easy to, use. Thcy are designed to solve particular sorts of problems and are there fore described as 'problem-orientated'.

It is also possible to write source programs in a low-level language. These languages resemble machine code more closely than natural language and are therefore difficult to, use. They art designed for particular computers and are thus described as 'machine-orientated'. Assembly lan-guages are low-level lan-guages. A program written in an assembly language is converted into machine code by means of a special program known as an assembler.

projectile: An object falling freely in a gravitational field, having been projected at a speed v and at an angle of elevation θ to, the horizontal. In the special case that θ = 90° the motion is linear in the vertical direction. If may then be treated using the equations of motion. tIn ail other cases the vertical and horizontal components of velocity must be treated separately. In the absence of friction, the horizontal component is constant and the vertical motion may be treated using the equations of motion. The path of the projectile is an arc of a parabola. Some useful relations are given below:

Time to reach maximum height:

$$t = v\sin\theta/g$$

Maximum height:

$$h = v^2 \sin^2\theta/g$$

Horizontal range:

$$R = v^2\sin^2 \theta/g$$

projection: A geometrical transformation in which one fine, shape, etc., is converted into, another according to certain geometrical rules. A set of points (the object) is converted into another set (the image) by the projection.

projective geometry: The study of how the geometric properties of a figure are altered by projection There is a one-to one correspon-dence between points in a figure and its projected image, but often the ratios of lengths will he changed. In central projection for example, a triangle maps into a triangle and a quadrilateral into a quadrilateral, but the sides and angles may change.

prolate: Denoting a spheroid that has a polar diameter that is greater than the equatorial diameter.

proof: A logical argument showing that a statement, proposition, or mathematical formula is true. A proof consists of a set of basic assumptions, called axioms or premisses, that are combined according. In logical rules, to derive as a conclusion the formula that is being proved. A proof of a proposition or formula P is just a valid argument from, true premisses to P as a conclusion.

proposition: A sentence or formula in a logical argument. A proposition can have a truth value; that is, it can be either true or false but not both. Any logical argument consists of a succession of propositions linked by logical operations with a proposition as conclusion. Propositions may be simple or compound. A compound proposition is one that is made up of more than one proposition. For example, a proposition P might consist of the constituent parts if R, then S or not Q; i.e. in this case $P = R \rightarrow (S \vee Q)$. A simple proposition in one that is not compound.

protractor: A drawing instrument used for marking out or measuring angles. It usually consists of a semicircular piece of transparent plastic sheet, marked with radial lines at one degree intervals.

pulley: A class of machine. In any pulley system power is transferred through the tension in a string wound river one or more wheels. The force ratio and distance ratio depend on the relative arrangement of strings and wheels. The efficiency is not usually very high as work must be done to overcome friction in the strings and the wheel bearings and to lift any moving wheels.

pure mathematics: The study of mathematical theory and structures without necessa-rily having an immediate application in mind. For example the study of the general of vectors considered purely as entities with certain properties certain properties, could be considered as a branch pure mathematics. The use of vector algebra in mechanics to solve a problem on forces or relative velocity is a branch of applied mathematics. Pure mathe-matics, then, deals with abstract entities, without any necessary reference to, physical applications in the 'real world'.

Q

quadrant: 1. One of four divisions of a plane. In rectangular Cartesian coordi-nates, the first a quadrant is the area to the right of the y-axis, and above the x-axis is, where both x and y are positive. The second quadrantes the area to the left the and above the where x is negative and y is positive. In polar coordinates, the first second third and fourth quadrarts occur when the direction angle, q, is 0 to 90°; 90° to 180°; 180° to 270° adn 270° to 360°, respectively. 2. A quarter of a circle, bounded by two perpendicular radii and a quarter of the circumference. 3. A unit of plane angle equal to 90° degrees ($\pi/2$ radians). A quadrant is a right angle.

quadrilateral: A plane figure with four straight sides. For example, squares, kites, rhombuses, and trapeziums are all quadrilaterals. A square is a regular quadri-lateral.

quart: A unit of capacity equal to two UK pints (in the UK) or two US pints (in the USA). In the USA a dry quart is equal to two US dry pints.

quartic equation: A poly-nomial equationin which the highest power of the unknown variable is four. The general form of a quartic equation in a variable x is

$$ax^4 + bx^3 + cx^2 + dx + e = 0$$

where a, b, c, d and e.are constants. Its also sometimes written in the reduced form

$$x^4 + bx^3/a + cx^2/a + dx/a + e/a = 0$$

In general, there are four values of x that satisfy a quartic equation. For example,

$$2x^4 + 9x^3 + 4x^2 + 21x - 18 = 0$$

can be factorized to

$$(2x+3)(x-1)(x-2)(x \quad 3) = 0$$

and its solutions (or roots) are –3/2, 1, 2, and 3. On a Cartesian coordinategraph, the curve

$$y = 2x^4 - 9x^3 + 4x^2 + 21x = 18 = 0$$

crosses the x-axis at $x = 3/2$; $x = 1$; and $x = 2$; $x = 3$.

quartile: One of the three points that divide a set of data arranged in numerical order into four equal parts. The lower (or first) quartile, Q_1, is the 25th percentile (P_{25}) The middle (or second) quartile, Q_2, is the median, The upper (or third) quartile, Q, is the 75th percentile (P_{15}).

Quotient: The result of dividing one number by another. There may or not be a remainder.

R

radius: The distance from the centre of a circle to any point on its circumference or from the centre of a sphere to its surface. In polar coordinates, a radius r (distance from a fixed origin) is used with angular position θ to specify the positions of points.

radius of gyration: Symbol k For a body of mass *m* and moment of inertia I about an axis, the radius of gyration about that axis is given by

$$k^2 = I/m.$$

In other words, a point mass *m* rotating at a distance *k* from the axis would have the same moment of inertia as the body.

radius vector: The vector that represents the distance and direction of a point from the origin in a polar coordinate system.

random access: A method of organizing information in a computer storage device so that the piece of information may be reached directly in about the same lime as any other. Main store, disk units, and drum units all operate by random access and are thus known as random access-memory (RAM). In contrast a magnetic tape unit operates more slowly by serial access: a particular piece of information can only be retrieved by working through the preceding blocks of data on the tape.

random number table: A table consisting of a sequence of randomly chosen digits from 0 to 9, where each digit has a probability of 0.1 of appearing in a particular position, and choices for different positions are independent. In random sampling from a population of size *n* every individual can be assigned a different number from 1 to *n*. If *k* is the number of digits *n*, the digits in the table are formed into groups of size *k*. The numbers so formed are read off, those greater than n being rejected, and matched with individuals until a sample of suitable size has been chosen.

random variable: (chance variable, stochastic variable) A quantity that can take any one of a number of unpredicted values. A discrete random variable, X, has a definite set of possible values $x_1, x_2, x_3...x_m$ with corresponding probabilities $p_1, p_2, p_3,...$ Since X must take one of the value in this set, $p_1 + p_2 + ... p_n = 1$

If X is a *continuous random variable*, it can take any value over a continuous range. The probabilities of a particular value x occurring is given by a *probability density function* $f(x)$. On a graph of $f(x)$ against x, the area under the curve between two values a and b is the probability that X lies between a and b. The total area under the curve is 1.

range: 1. The difference between the largest and smallest values in a set of data. It is a measure of dispersion. In terms of percentiles, the range is $(P_{100} - P_0)$. 2. A set of numbers or quantities that form possible results of a mapping. In algebra, the range of a function $f(x)$ is the set of values that $f(x)$ can take, for ail possible values of x. For example, if $f(x)$ is taking the square root of positive rational numbers, then the range would be the set of real numbers.

rank: A method of ordering a set of objects according to, the magnitude or importance of a variable measured on them, e.g., arranging ten men in order of height. If the objects are ranked using two different variables, the degree of association between the two rankings is given by the coefficient of tank correla-tion.

rates: A tax paid to local government authorities by the owners of property. Each property is given a rateable value, which is assessed by a valuation officer. Each year the local authority deciares its rate poundage, i.e. the amount that each ratepayer must pay for each Pound of rateable value of his property. Thus the amount of rates paid is determined each year by multiplying the rateable value by the rate poundage.

ratio: One number or quantity divided by another. The ratio of two variable quantities x and y, written as x/y or $x{:}y$, is constant if y is proportional to x.

rationalized units: A system of units in which the equations have a logical form related to the shape of the system. SI units form a rationalized system of units. For example, in it formulae concerned with circular symmetry constant a factor of 2π; those concerned with radial symmetry contain a factor of 4π.

rational numbers: Symbol: Q The set of numbers that includes integers and fractions. Rational numbers can be written down exactly as ratios or as finite or repeating decimals.

reaction: Newton's third law of force states that whenever object A applies a for, e on object B, B applies the same force on A. An old word for force is 'action'; 'reaction' is thus the other member of the pair.

Often it is hard or impossible, to decide what A and B are. Thus in the interaction between two electric charges, each exerts a force on the other. Thus, in general, action and reaction have little meaning. The word 'reaction' is still sometimes used in restricted cases, such as the reaction of a support on the object it supports. In the case the 'action' is the effect of the weight of the object on the support.

reader: A device used in a computer system to sense the information recorded on some source and convect in into another form. A paper-tape reader, for example, senses the series of holes punched in a paper tape and converts the information into a series of electrical pulses, which can be transmitted to the central processor of the computer.

real numbers: Syrnbol Q The set of numbers that includes all rational and irrational numbers.

real time: The actual time in which a physical process takes place or in which a physical process, machine, etc., is under the direct control of a computer. A real time system is able to react sufficiently rapidly so that it may control a continuing process, making changes or modifications when neces-sary. Air traffic control and airline reservations require realtime systems.

reciprocal: The number 1 divided by a quantity, For example, the reciprocal of 2 is 1/2. The reciprocal of (x^2+1) is $1/(x^2+1)$. The product of any expression and its reciprocal is 1. For any function, the reciprocal is the multiplicative inverse.

rectangle: A plane figure with four straight sides, two parallel pairs of equal length forming right angles. The area of a rectangle is the product of the two different side lengths, the length times the breadth. A rectangle has two axes of symmetry, the two lines joining the midpoints of opposite sides. It can also be superimposed on itself after rotation through 180° (π radians). The two diagonals of a rectangle have equal lengths.

reduced form: (of a polynomial) The equation of the form

$$x^n + (b/a)x^{n-1} + (c/a)x^{n2} + \dots = 0$$

that is derived from a polynomial of the form

$$ax^2 + bx^{n-1} + cx^{n-2} + ... = 0$$

For example,

$$2x^2 - 9x + 12 = 0$$

is equivalent to the reduced form

$$x^2 - 9x/ \cdot 2 + 6 = 0$$

Both equations have the solution x 3,12 and $x = 4$. The Cartesian coordinate graph of

$$y = 2x^2 - 9x + 12$$

crosses the x-axis at the same points, $x = 3/2$ and $x = 4$ as the graph of

$$y = x^2 - 9x/2 + 6$$

reduction formulae: In trigonometry the equations that express the sine, cosene and tangent functions of an angle in terms of an angle between 0 and 90° ($\pi/2$).

reflection: The geometrical transformation of a point or a set of points from one side of a point, line, or plane to a symmetrical position on the other side. On *reflection in a line,* the image of a point P would be a point P′ at the same distance from the line but on the other side. The line, the *axis of reflection,* is dit perpendicular bisector of the line PP′. In a symmetrical plane figure there is an axis of reflection, also called the axis of symmetry, in which the figure is reflected onto itself. An equilateral triangle, for example, has the axes of symmetry. In a circle, any diameter is an axis of symmetry. Similarly a solid may undergo reflection in a plane. In a sphere, any plane passing through the sphere s centre would be a plane of symmetry.

In a Cartesian coordinate system, reflection in the x-axis changes the sign of the y-coordinate. A point (a, b) would become $(a,-b)$. Reflection in the changes the y-axis changes the sign of z-coordinate is equivalent to reflection in the plane of the x and y axes. *Reflection in a point* is equivalent to rotation through 180°. Each point P is moved to a position P′ so that the point of reflection bisects the line PP′. Reflection in the origin of Cartesian coordinates changes the signs of all the coordinates. It is equivalent to reflection in the x-axis followed by reflection in the y-axis, or vice versa.

regression line: A line $y = ax + b$, called the regression line of y on x, which gives the expected value of random variable y conditional on the given value of a random variable x. The regression line of x on y is not in general the same as that of y on x. If a diagram of a data point $(x_1, y_1),...(x_m, y_n)$ is drawn by hand a linear relationship is shown up the line can be drawn by hand. The best line is drawn using the least squares method.

relative: Expressed as a difference from or as a ratio to, some reference level. Relative density, for example, is the mass of a substance per unit volume expressed as a fraction of a standard density, such a that of water.

relative error: The error or uncertainty in a measure-ment expressed as a fraction of the measurement. For example, if, in measuring a length of 10 metres, the tape measures only to the nearest centimetre, then the measurement might be written as 10 ± 0.01 metres. The relative error is 0.0/10 = 0.001.

relative velocity: If two objects are moving at velocities v_A and v_B in a given direction, the velocity of A relative to B is v_A v_B in that direction. In general, if two objects are involving in the same frame at nonrelativistic speeds their relative velocity is the vector difference of the two velocities.

relativistic mass: The mass of an object as measured by an observer at test in a frame of reference in which the object is moving with a velocity v. It is given by

$$m_0 = m\surd\, 1 - v^2/c^2$$

where m_0 is the rest mass, c is the velocity of light, and in is the relativistic mass. That equation is a consequence of the special theory of relativity, and is in excellent agreement with experiment. No object can travel at the speed of light because its mass would then be infinite.

relativistic mechanics: A system of mechanics based on relativity theory.

relativisti^ speed: (relativistic velocity) Any speed (velocity) that is sufficiently high to make the mass of an object significantly greater than its test mass. It is usually expressed as a fraction of c, the speed of light in free space. At a speed of c/2 the relativistic mass of an object is about 15% greater than the test mass.

relativity theory: A theory put forward in two parts by Albert Einstein. The special theory referred only to non-accelerated (inertial) frames of reference. The general theory is also applicable to accelerated systems.

The special theory was based on two postulates: (1) That physical laws are the same in all inertial frames of reference. (2) That the speed of light in a vacuum is constant for all observers, regardless , of the motion of the source or observer. The second postulates seems contrary to 'common sense' ideas of motion. Einstein was led to the theory by considering the problem of the 'ether' and the relation between electric and magnetic fields in relative motion. The theory accounts for the negative result of the Michelson-Morley experi-ment and shows that the Lorentz-Fitzgerald contrac-tion is only an apparent effect of motion of an object relative to an observer, net a 'real' contraction. It leads to the result that the mass of an object moving ai a speed v relative to an observer is given by:

$$m = m_0 \sqrt{(1 - v^2/c^2)}$$

where c is the speed of light and in the mass of the object when at rest relative to the observer. The increase in mass is significant at high speeds. Another consequence of the theory is that an object has an energy content by virtue of its mass, and similarly that energy has inertia. Mass and energy are related by the famous equation $E = mc^2$

The general theory of relativity seeks to explain the difference between accelera-ted and non-accelerated systems and the nature of the forces acting in both of them. For example, a person in a spacecraft fat out in space would not be subject to gravitational forces. If the craft were rotating, he would be pressed against the walls, of the craft and would consider that he had weight. There would not be any difference between this force and the force of gravity. To an outside observer the force is simply a result of the tendency to continue in a straight line; i.e. his inertia. This type of analysis of forces led Einstein to a 42 principle of equivalence that inertial forces and gravitational forces are equivalent, and that gravitation can be explained by the geometrical properties of space. He visualised a four-dimensional space-time continuum in which the presence of a mass affects the geometry the space is 'curved' by the mass.

remainder: The number left when one number is divided into another. Dividing 12 into 57 gives 4 remainder 9 (4 x 12 = 48; 57– 48 = 9),

remainder theorem: The theorem expressed by the equation

$$f(x) = (\ x - a\ )g(x) + f(a).$$

This means that if a polynomial in x, $f(x)$, is divided by $(x - a)$, where a is a constant, the remainder term is equal to the value of the polynomial when $x–a$. For example, if

$$2x^3 + 3x^2 - x - 4$$

is divided by $(x - 4)$, then the remainder term is

$$f(4) = 128 + 48 - 4 - 4 = 168$$

The remainder theorem is useful for finding the factors of a polynomial. In this example:

$$f(1) = 2 + 3 - 1 - 4 = 0$$

so $(x - 1)$ is a factor.

representative fraction: A fraction used to express the scale of a map in which the numerator represents a distance on the map and the denominator represents the corresponding distance on the ground. As a fraction is a ratio, the units of the numerator and denominator must be the same. For example, a scale of 1 cm = 1 km would be given as a representative fraction of 1/100000, because there are 100000 cm in 1 km.

resolution of vectors: The determination of the components of a vector in two given directions ai 90°. The term is sometimes used in relation to finding 14 any pair of components.

resonance: The large-amplitude vibrations of in object or system when given impulses at us natural frequency. For instance, a pendulum swing with a natural frequency that depends on its length and mass. If it is given a periodic 'push' at this frequency for example, at each maximum of a complete oscillation— the amplitude is increased with little effort. Much more effort would be required to produce a swing of the same amplitude, at a different frequency.

restitution coefficient of Symbol: e For the impact of two, bodies, the elasticity of the collision is measured by the coefficient of restitution. It is the relative velocity alter collision divided by the relative velocity before collision (velocities measured along the line of centres). For spheres A and B

$$v_A' - v_B' = e(v_A - v_B)$$

v indicates velocity before collision; v′ velocity after collision. Kinetic energy is conserved only in a perfectly elastic collision.

rest mass: Symbol: m_0 The mass of an object at test as measured by an observer at test in the same frame of reference.

resultant: A vector with the same effect as a number of vectors. Thus, the resultant of a set of forces is a force that has the same effect; it is equal in magnitude and opposite in direction to the equilibrant. Depending on the circumstances, the resultant of a set of vectors can be found by different methods.

rhombohedron: A solid figure bounded by six faces, each one a parallelogram, with opposite faces congruent.

rhomboid: A parallelogram that is neither 3 a rhombus nor a rectangle.

rhombus: A plane figure with four straight sides of equal length. Its area is equal to half the product of the lengths of its two diagonals, which bisect each other perpendicularly. The rhombus is symmetrical about both of its diagonals and also has rotational symmetry, in that it can be superimposed on itself after rotation through 180°.

Riemann sum: The series that approximates the area between the curve of a function f(x) and the x-axis:

$$f(\xi_i)\Delta x_i$$

where x is an increment of x, j is any value of f(x) within that interval, and n is the number of intervals. The definite (or Riemann) integral is the limit of the sum as n becomes infinitely large and Δx infinitesimally small.

right angle: An angle that is 90° or $\pi/2$ radians. It is the angle between two lines or planes that are perpendicular to each other. The corner of a square, for example, is a right angle.

right solid: A solid geometrical figure that is upright; for example, a cone, cylinder, pyramid, or prism that has an axis at right angles to the base.

rigid body: In mechanics, a body for which any change of shape produced by forces on the body can be neglected in the calculations.

Rolle's theorem: A curve that interacts the x-axis at two points a and b, is continuous, and has a tangent at every point between a and b, must have at least one point in this interval at winch the tangent to the curve is horizontal. For a curve y = f(x), it follows from Rolle's theorem that the function f(x) has a turning point (a maximum or minimum value) between f(a) and f(b), where the derivative f(x) = 0.

root: In an equation, a value of the independent variable that satisfies the equation. In general, the degree of a polynomial is equal to the number of room, A quadratic equation (one of degree two) has two roofs, although in same circumstances they may be equal. For a number a, an nth root of a number that satisfies the equation

$$x^n = a$$

rose: A curve obtained by plotting the equation

$$r = a\sin\ n\theta$$

in polar coordinates (a is a real-number constant and n is an integer constant). It has a number of petal-shaped loops, or leafs. When n is even there are $2n$ loops and when n is odd there are n loops. For example, the graph of $r = a\sin 2\theta$ is a four-leafed rose.

rotation: A geometrical transformation in which a figure is moved rigidly around a fixed point. If the point, the centre of rotation, is labelled O, then for any point P in the figure moving to point P′ after rotation, the angle POP′ is the same. This angle is the angle of rotation. Some figures are unchanged by certain rotations, A circle is not affected by any rotation about its centre. A square does not change it is rotated through 90° about the point at which its diagonals cross. Similarly an equilateral triangle is unchanged by rotation through 120° about its centroid. The properties are known as the rotational symmetry of the figure.

rotational motion: Motion of a body turning about an axis. The physical quantities and laws used to describe linear motion ail have rotational 44 analogues: the equations of rotational motion are the analogues of the equations of motion (linear).

rotation of axes: In coordinate geometry, the shifting of the reference axes so that they are rotated with respect to the original axes, of the system by angle (θ). If the new axes are x' and y' and the original axes x and y, then the coordinates (x, y) of a point with the original axes are related to the new coordinates (x', y') by:

$$x = x' \cos\theta - y' \sin\theta$$

$$y = x' \sin\theta - y' \cos\theta$$

rough: In mechanics, describing a system in which frictional effects have to be taken into consideration in the calculations,

rounding: (rounding off) The process of adjusting the least significant digit or digits in a number after a required number of digits has been truncated (dropped), This reduces the error arising from truncation but, still leaves a rounding error so that the accuracy of, say the result of a calculation will be decreased. For example, the number 2.871 329 71 could be truncated to 2.87132 but would be rounded to 2.871 33.

routine: A sequence of instructions used in computer programming. It may be a short program or sometimes part of a program.

row vector: (row matrix) A number, (n), of quantities arranged in a row; i.e. a $1 \times n$ matrix. For example, the coordinates of a point in a Cartesian coordinate system with three axes is a 1×3 row vector, (x, y, z).

Runge-Kutta method: An iterative technique for solving ordinary differential equations, used in computer analysis.

S

Sampling: The selection of a representative subset from a whole population. Analysis of the sample gives information about the whole population. This is called statistical inference. For example population parameters can be estimated using sample statistics (such as the sample mean and variance). Significance (or hypothesis) tests are used to test whether observed differences between two samples are due to chance variation or are significant, as in testing a new production process against an old. The population can be finite or infinite. In sampling with replacement each individual chosen is returned to the population before the next choice is made. In random sampling every member of the population has an equal chance of being chosen. In stratified random sampling the population is ordered, the first individual chosen at random and further individuals chosen at specified intervals, for example, every 100th person on the electoral roll. If a random sample or size n is the set of numerical values $(x_1, x_2, ..., x_n)$, the sample mean is:

$$x = x_i/n$$

The sample variance is:

$$\Sigma(x_i - x)^2/(n - 1)$$

or $$\Sigma(x_i - x)^2/n$$

for a normal distribution If μ is the population mean, the sample variance is:

$$\Sigma(x_i - \mu)/n$$

Sampling distribution: The distribution of a sample statistic. For example, when different samples of size n arc taken from the same population the means of each sample form a sampling distribution. If the population is infinite (or very large) and sampling is with replacement, the mean of the sample means is ue u and the standard deviation of the sample means is $\mu_x = \mu$ In, where μ and α are the population mean and standard deviation. When $n \geq 30$ sampling distributions are approximately normal and large sampling theory is used. When $n < 30$ exact sample theory is used.

satisfy: To be a solution of. For example, 3 and –3 satisfy the equation $x^2 = 9$.

scalar: A number or a measure in which direction is unimportant or meaningless. For instance, distance is a scalar quantity, whereas displacement is a vector. Mass, temperature, and time are scalars–they are each quoted as a pure number with a nuit.

scalar product: A multiplica-tion of two vectors to give a scalar. The scalar product of A and B is defined by $A.B = AB \cos\theta$, where A and B are the magnitudes of A and B and θ is the angle between the vectors. An example is a force F displaced s. Here the scalar product is energy transferred.

$$W = F.s$$

$$W = Fs\cos\theta$$

where θ is the angle between the line of action of the force and the displacement. A scalar, product is, indicated by a dot between the vectors and is sometimes called a dot product. The scalar product is commutative

$$A.B = B.A$$

and is distributive with respect to vector addition

$$A.(B + C) = A.B + A.C$$

If A is perpendicular to B, A.B = 0. In two-dimensional Cartesian coordinates with unit vectors i and j in the x and y-directions respectively,

$$A.B = (a_1i+a_2j).(b_1i+b_2j) = a_1b_1+a_2b_2.$$

Scalar projection: The length of an orthogonal projection of one vector on another. For example, the scalar projection of A on B is $A \cos\theta$, or $(A.B)/b$ where θ is the smaller angle between A and B, and b is the unit vector in the direction of a B.

Scale: 1. The makings on the axes of a graph, or on a measuring instrument, that correspond to values of a quantity. Each unit of length or, a linear scale represents the same interval. For example, a thermometer that has markings 1 millimetre apart to represent 1°C temperature intervals has a linear scale

2. The ratio of the length of a line between two points on a map to the distance represented. For example, a map in which two points 5 kilometres apart are shown 5 centimetres apart has a scale of 1/100000.

scale factor: The multiplying factor for each linear measurement of an object when it is to be enlarged about a given centre of enlargement. A scale factor can be positive, negative, or fractional. If the scaler factor is positive, the image is larger than the object and on the same side of the centre of enlargement. If the scale factor is negative, the image will be smaller than object but on the same side of the centre of enlargement.

Scalene: 1. Denoting a triangle with three unequal sides.

2. Denoting a cone or cylinder that has an axis not at right-angles to the base.

Scatter diagram: (Galton graph) A graphical representation of data from a bivariate distribution (x,y). The variables are measured on n individuals giving data (x_2y_1),... (x_n,y_n) e.g., X_1 and Y_i are the height and weight of the eh individual. If Y_1 is plotted against X_1 the resultant scatter diagram will indicate any relationship between x and y by showing if a smooth curve can be drawn through the points.

scientific notation: (standard form) A number written as the product of a number between 1 and 10 with a power of 10. For example, 2342.6 in scientific notation, n is 2.3426×10^3, and 0.0042 is written as 4.2×10^{-3}

screw: A type of machine, related to the inclined plane and, in practise to the second order lever. The efficiency of screw system is very low because of friction Even so the force ratio (F_2/F_1)can be very high.

The distance ratio is given by $2\pi r/p$, where r is the radius and p the pitch of the, screw.

secant: 1. A line that intersects a curve. The intercept is a chord of the curve.

2. A trigonometric function of an angle equal to the reciprocal of its cosine; i.e., $\sec\alpha = 1/\cos\alpha$.

second order differential equation: A differential equation in which the highest derivative of the dependent variable is a second derivative

sector: part of a circle formed between two radii and the circumference. The area of a sector is $\frac{1}{2}r^2\theta$, where r is the radius and θ is the angle in radians, formed at the centre of the, circle between the two radii.

Segment: part of a line or curve between two points, part of a plane figure but off by a straight line, or part of a solid cutoff by a Plane. For example, on a graph, a line segment may show the values of a function within a certain range. The area between a chord of a circle and the corresponding arc is a segment of the circle. A cut through a cube parallel to one of ifs face, forms two cuboid segments.

semicircle Hall: a circle, bounded by a diameter and half of the circumferences

separation of variables: A method of solving ordinary differential equations. In a first order differential equation,

$$dy/dx = F(x,y)$$

if $F(x,y)$ can be written as $f(x)$, $g(y)$, the variables in the function are separable and the equation can therefore be solved by writing if as

$$dy/g(y) = f(x)dx$$

and integrating both sides.

Sequence: (Progression) An ordered set of numbers. Each term in a sequence can be written as an algebraic function of its position. For example, in the sequence {2, 4, 6, 8, ... } the general expression for the nth term is $a_n = 2n$. A finite sequence has an infinite number of terms.

series: The sum of an ordered set of numbers. Each term in the series can be written as an algebraic function of its position. For example in the series 2 +4 + 6 + 8 + ... the general expression for the nth term, a_n is $2n$. A finite series has a finite fixed number of terms. An infinite series has an infinite number of terms. A series with m terms, or the sum of the first m terms of an infinite series, can be written as S_m or

$$\Sigma a_n$$

Set: Any collection of things or numbers that belong to a well-defined category. For example, 'dog' is a member, or element, of the set of 'types of four-legged animal'. The set of 'days in the week' has seven elements. In set notation, this would be written as n{Monday, Tuesday, ... } = 7. This kind of set is a finite set. Some sets, such as the set of natural numbers, N = {l, 2, 3, ... }, have an infinite number of elements. A line segment also is an infinite set of points.

Another way of writing a set of numbers is by defining it algebraically. The set of all numbers between 0 and 10 could be written as $\{x : 0 < x < 10)$. That is allvalues of a variable, x, such that x is greater than zero and less than ten.

Set square: A drawing instrument, consisting of a flat right-angled triangular shape, used for drawing right angles and angles of 30°, 45°, and 60°. Some types have a variable scale so that other angles can be drawn.

Sexagesimal: Based on multiples of 60. The measurement of an angle in degrees, minutes, and seconds, for example, is asexagesimal measure, because there are 60 seconds in one minute, and 60 minutes in one degree. A sexagesimal number is one that uses 60 as a base instead of 10.

Significant figure: The number of digits used to, denote an exact value to a specified degree of accuracy. For example, 6084.324 is a value accurate to seven significant figures. If it is written as approximately 6080, it is accurate to three significant figures. The final 0 is not significant because it is used only to show the order of magnitude of the number.

Similar: Denoting two Or more figures that differ in scale but not in shape. The conditions for two triangles to be similar are:

(1) Three sides of one are proportional to three sides of the other.

(2) The angle of one is equal to the angle of the other and the sides; forming the angle in one proportional to the same sides in the other.

(3) Three angles of one are equal to three angles of the other

simple harmonic motion: (s.h.m.) Any motion that can be drawn as a sine wave. Examples are the simple oscillation (vibration) of a pendulum or a sound source and the variation involved in a simple wave motion. Simple harmonic nation is observed when the system, moved away from the central position, experiences a restoring force that is proportional to, the displacement from ibis Position.

The equation of motion for such a system can be written in the form

$$md^2x/dt^2 = -\lambda x$$

λ being a constant. During the motion there is an exchange of kinetic and potential energy, the sum of the two being constant (in the absence of damping). The period (T) is given by

$$T = 1/f$$

$$T = 2\pi/\omega$$

where f is frequency and ω pulsatance. Other relation-ships are:

$$x = x_0 \sin \omega t$$

$$dx/dt = \pm\ \omega\sqrt{(x_0^2 - x^2)}$$

$$d^2x/dt^2 = -w^2x$$

Here x_0 is the maximum displacement, i.e. the amplitude of the vibration. In the case of angular motion, as for a pendulum, θ is used rather than x. A simple harmonic motion can be represented by the motion of a point at constant speed in a circular path. The perpendicular from the point be an axis through a diameter describes a simple harmonic motion This is used in a method of representing simple harmonic motions by rotating vectors.

simple interest: The interest earned on capital, when the interest is withdrawn as it is paid, so that the capital remains fixed. If the amount of money invested (the principal) is denoted by P, the time in years by T and the rate per annum in percent by R, then the simple interest is $PRT/100$.

Simpson's rule: A rule for finding the approximate area under a curve by dividing it into pairs of vertical columns of equal width with bases lying along the horizontal axis. Each pair of columns is bounded by the vertical lines from the x-axis be corresponding points on the curve and at the top by a parabola that goes through these three point which approximates the curve. For example, if the value of f(x) is known at $x = a$, $x = b$, and at a value mid-way between a and b the integral of f(x)dx between limits a and b is approximately equal to: $h/3(f(a) + 4f[(a + b)/2] + f(b))$

h is hall the distance between a and b. As with the trapezium rule, which is less accurate, better approxima-tions can be obtained by subdividing the area into 4,6, 8, ... columns will further subdivision makes no significant difference to the result.

simultaneous equations: A set of two or more equations that together specify conditions for two or more variables. If the number of unknown variables is the same as the number of equations, then there is a unique value for each variable, which satisfies all the equations. For example, the equations

$$x + 2y = 6$$

and

$$3x + 4y = 9$$

have the solution $x = -3$; $y = -1.5$. The method of solving simultaneous; equations is to eliminate one of the variables by adding or subtracting the equations. For example, multiplying the first equation, above by 2 and subtracting it from the second gives:

$$3x + 4y - 2x - 4y = 9 - 12$$

i.e. $x = -s3$. Substituting this into either equation gives the value of y. Simultaneous equations can also be solved graphically. On a Cartesian coordinate graph, each equation would be shown as a straight fine and the point at which the two cross is in this case (–3, –1.5).

Sine: (sin) A trigonometric function of an angle. The sine of an angle a (stria) in a right-angled triangle is the ratio of the side opposite the angle to the hypotenuse. This definition applies only to angles between 0° and 90°.

More generally, in rectangular Cartesian coordinates, the y-coordinate of any point on the circumference of a circle of radius r centred at the origin is rsina, where a is the angle between the x-axis and the radius to that point. In other words, the sine function depends on the vertical component of a point on a circle. Sina is zero when α is 0°, rises to 1 when $\alpha = 90°$ ($\pi/2$) falls again to zero when , a 180°(π), becomes negative and reaches -1 at $\alpha = 270°(3\pi/2)$, and then returns to zero at a =360° (2π). This cycle is repeated every complete revolution. The sine function has the following properties:

$$\sin\alpha = \sin(\alpha + 360°)$$

sine wave: The waveform resulting from plotting the sine of an angle against the angle. Any motion that can be plotted so as to give a sine wave is a simple harmonic motion.

Singular matrix: A square matrix for which the determinant is zero and that has no inverse matrix.

Singular point: A point on a curve y f(x) at which the derivative dy/dx has the indeterminate form, 0/0. The singular points on a curve are found by writing the derivative in the form

$$dy/dx = g(x)/h(x)$$

and then finding the values of x for which $g(x)$ and $h(x)$ are both zero.

SI units: (System International d'Unites) The internationally adopted system of units used for scientific purposes. It has seven base units (the metre, kilogram, second, kelvin, ampere, mole, and candela) and two supplementary units (the radian and steradian). Derived units are formed by multiplication and/or division of base units; a number have special names. Standard prefixes are used for multiples and submultiples of SI units. The SI system is a coherent rationalized system of units.

Skewness: A measure of the degree of a symmetry of a distribution. If the frequency curve has a long tail to the right (left) and a short one to the left (right), it is called skewed to the right (left) or positively (negatively) skewed. Skewness is measured by either Pearson's first measure of skewness, (mean mode) divided by standard deviation, or the equivalent Pearson's second measure of skewness, divided by standard deviation.

Slant height: 1. The length of an element of a right cone; i.e. the distance from the vertex to the directrix.

2. The altitude of the faces of a right pyramid.

Slide rule: A calculating device on which logarithmic scales are used to multiply numbers. Most slide rules also have fixed scales showing squares, cubes and trigonometric functions. The accuracy of the slide rule is usually to three significant figures. A scale is marked across the join between a sliding middle section and a fixed outer part. To multiply two numbers, for example, 2.1 x 3.2, the zero on the middle section is lined up with 2.1 on the outer scale. Lined up with 3.2 on the inner scale with the product 3.2 × 2.1.

Smooth: In mechanics, describing a system in which friction can be neglected in the calculations.

Software: The programs that can be run on a computer, together with any associated written documentation. A software package is a professionally written program or group of programs that is designed to, perform some commonly required task, such as statistics or graph plotting, and is fully documented. The availability of software packages means that common tasks need not be programmed over and over again.

Solid angle: Symbol Ω The three-dimensional analogue of angle; the region subtended at a point by a surface (rather than by a fine). The unit is the steradian (sr), which is defined analogously to the radian the solid angle subtending unit area at unit distance. As the area of a sphere is $4\pi r^2$. the solid angle corresponding to the revolution (2π radians) is 4π steradians.

Solid of revolution: A solid figure that can be produced by revolution of a fine or curve the generator) about a fixed axis. For instance, rotating a circle about a diameter generates a sphere. Rotating a circle about an axis that does not cut the circle generates a torus

solution: A value of a variable that satisfies an algebraic equation. For example, the solution of $2x + 4 = 12$ is $x = 4$. An equation may have more than one solution; for example, $x^2 = 16$ has two; $x = -4$ and $x = +4$.

Solution of triangles: Calculating the unknown sides and angles in triangles. Because the suer of angles in a triangles always 180, the third angle can be calculated if two are known. All the sides and angles can be calculated when two sides and the angle between them are, known, but if two sides and another angle are known there are two possible solutions. Any two angles and one side are sufficient to solve a triangle.

Space curve: A curved fine in a volume, defined in three-dimensional Cartesian coordinates by three functions:

$$x = f(t)$$

$$y = g(t))$$

$$z = h(t)$$

or by two equations of the form:

$$F(x,y,z) = 0$$

$$G(x,y,z) = 0$$

space-time: In Newtonian (pre-relativity) physics space and time are separate and absolute quantifies; that is they are the same for all observers in any frame of reference. An event seen in one frame is also seen in the same place and at the same time by another observer in a different frame.

After Einstein had proposed his theory of relativity, Minkowski suggested that since space and time could no longer be regarded as separate continua, they should be replaced by a single continuum of four dimensions, called space-time. In space-time the history of an object's motion in the course of time is represented by a line called the world curve.

Spearman's method: A way of measuring the degree of association between two rankings of n objects using two different variables x and y which give data (x_1,y_1), ..., (x_m,y_n). The objects are ranked using first the x's and then the y's, and the difference, D, between the ranks calculated for each object.

Spearman's coefficient of rank correlation

$$p = 1 -(6\Sigma D^2/[n(n^2 - 1)])$$

Speed: Symbol c Distance moved per unit time: $c = d/t$. Speed is a scalar quantity; the equivalent vector is velocity a vector quantity equal to displacement per unit time.

Usage can be confusing and if is common to meet the word 'velocity' where 'speed' is more correct. For instance c_0 is the speed of light in free space, not its velocity.

Sphere: A closed surface consisting of the locus of points in space that are at a fixed distance, the radius r, from a fixed point, the centre. A sphere is generated when a circle is turned through a complete revolution about an axis that is one of its diameters. The plane cross-sections of a sphere are all circles. The sphere is symmetrical about any plane that passes through its centre and the two mirror image shapes on each side are called hemispheres. In Cartesian coordinates, the equation of a sphere of radius r with its centre at the origin is

$$x_2 + y_2 + z_2 = r_2$$

The volume of a sphere is $4\pi r^3/3$ and its surface area is $4\pi r^2$

spherical polar coordinates: A method of defining the position of a point in space by its radial distance, r, from a fixed point, the origin 0, and its angular position on the surface of a sphere centred at O. The angular position is given as two angles θ and ϕ . θ is the angle that the radius vector makes with a vertical axis through O (from the south pole to the north pole). It is called the colatitude. For points on the vertical axis above O, $\theta = 0$. For points lying in the 'equatorial' horizontal plane, $\theta = 90°$. For points on the vertical axis below O, $\theta = 180°$. ϕ is the angle that the radius vector makes with an axis in the equatorial plane. It is called the azimuth. For ail points lying in the axial plane, that is, on, vertically above, or vertically below this axis, $\phi = 0$ on the positive side of 0 and $\phi = 180°$ on the negative side. This plane corresponds to the plane $y = 0$ in rectangular Cartesian coordinates. For points in the vertical plane at 90° to this, ($x = 0$ in rectangular Cartesian coordinates) $\phi = 90°$ in the negative half-plane and 270° in the negative half-plane. For a point the corres-ponding rectangular Carte-sian coordinates (x,y,x) are:

$$x = r\cos\phi\sin\theta$$

$$y = r\sin\phi\sin\theta$$

$$z = r\cos\theta$$

Spherical sector: The solid generated by rotating a sector of a circle about a diameter of the circle. The volume of a sperical sector generated by sector of altitude h parallel to the axis of rotation) in circle of radius r is

$$(2/3\pi r^2 h$$

Spherical trigonometry: The study and solution of spherical triangles.

Spheroid: A body or curved surface that is similar to a sphere but is lengthened or shortened in one direction.

Square: 1. The second power of a number or variable. The square of x is $x\ x \times x = x^2$ (x squared). The square of the square root of a number is equal to that number 2. In geometry, a plane figure with four equal straight sides and right angles between the sides. Its area is the length of one of the sides squared. A square has four axes of symmetry–the two diagonals, which arc of equal length and bisect each perpendicularly, and to two lines joining the midpoint of opposite sides. It can be superimposed on itself after rotation through 90°.

square matrix: A matrix that has the same number of rows and columns, that is, a square array of numbers. The diagonal from the top left to the bottom right of a square matrix is called the leading diagonal (or principal diagonal). The sum of the elements in this diagonal is called the trace (or spur) of the matrix.

Squaring the circle: The attempt to construct a square that has the same area as a particular circle, using a ruler and compasses. An exact solution is impossible because there is no exact length for the edge, which is a multiple of the transcen-dental number $\sqrt{\pi}$.

Stability: A measure of how hard it is to displace an object or system from equilibrium. Three cases are met in statics differing in the effect on the centre of mass of a small displacement. They are:

(1) Stable equilibrium the system returns to its original state when the displacing force is removed.

(2) Unstable equilibrium the system moves away from the original state when displaced a small distance.

(3) Neutral equilibrium when displaced a small distance the system is at equilibrium in its new position.

An object's stability is improved by: (a) lowering the centre of mass, or (b) increasing the area of support; or by both.

Stable equilibrium: Equili-brium such that if the system is disturbed a little, there is a tendency for it to return to its original state.

Standard: Established as a reference.

1. Writing an equation in standard form enables comparison with other equations of the same type. For example,

$$x^2/4^2 + y^2/2^2 = 1$$

and

$$x^2/3^2 + y^2/5^2 = 1$$

are equations of hyperbolas in rectangular Cartesian coordinates, both written in standard form.

2. A standard measuring instrument is one against which other instruments are calibrated.

3. Standard form of a number.

standard form of a number: A number written as a number between 1 and 10 multiplied by a power of ten. For example, 0,000 326 and 42 567 are written as 3.25×10^{-4} and 4.257×10^5 respectively in standard form.

standard pressure: An internationally agreed value: a barometric height of 760 mmHg at 0°sC; 101 325 Pa (approximately 100 kPa).

This is sometimes called the atmosphere (used as a unit of pressure). The bar, used mainly when discussing the weather, is 100 kPa exactly.

standard temperature: An internationally agreed value for which many measure-ments are quoted. It is the melting temperature of water, 0°C.

Static pressure: The pressure an a surface due to a second solid surface or to a fluid that is not flowing.

statics: A branch of mechanics dealing with the forces on an object or in a system in equilibrium. In such cases there is no resultant force or torque and therefore no resultant acceleration.

Stationary point: A point on a curved line at which the slope of the tangent to the curve is zero. All turning points (maximum points and minimum points) are stationary points. In this case the slope of the tangent passes through zero and changes its sign. Some stationary points are net turning points. In such cases, the curve levels out and then continues to increase or decrease as before. At a stationary point the derivative dy/dx of $y = f(x)$ vanishes (is zero). At a maximum, the second derivative, d^2y/dx^2, is negative; at a minimum it is positive. At a horizontal inflection point the second derivative is zero. Net all inflection points have dy/dx.

At a stationary point on a curved surface representing a function of two variables, $f(x,y)$, the partial derivatives $\partial f/\partial x$ and $\partial f/\partial y$ are both zero. This May lie a maximum point, a minimum point, or a saddle point.

stationary wave: (standing wave) The interference effect resulting from two waves of the same type moving with the same frequency through the same region. The effect is most often caused when a wave is reflected back along its own path. The resulting interference pattern is a stationary wave pattern, Here some points always show maximum amplitude. They are called *antinodes* and nodes receptively. The distance between neighbou-ring node and antinode is a quarter of a wavelength.

statistics: The methods of planning experiments, obtaining data, analysing it, drawing conclusions from it, and making decisions on the basis of the analysis. In statistical inference con-clusions about a population are interfered from analysis of a sample. In descriptive statistics data is summarized but no inferences are made.

Steradian: Symbol sr The SI unit of solid angle. The surface of a sphere, for example, subtends a solid angle of 4π at its centre. The solid angle of a cone is the area intercepted by the cone on the surface of a sphere of unit radius.

Stereographic projection: A geometrical transformation of a sphere onto a plane. A point is taken on the surface of the sphere the pole of the projection. The projection of points on the sphere onto a plane is obtained by taking straight lines from the pole through the points, and continuing them to the plane. The plane taken does not pass through the pole and is perpendicular to the diameter of the sphere through the pole.

stochastic process: A process that generates a series of tandem values of a variable and builds up a particular statistical distribution front these. For example, the Poisson distribution can be built up by a stochastic process that starts with values taken from tables of random numbers.

Stocks and shares: Stocks in the U.K. are usually fixed interest securities issued by the government or a limited company to raise capital. Investors lend their own capital to the issuers of the stock in return for a fixed rate af interest, which usually paid twice a year. Stocks are usually issued in units of £100 and their market price is largely determined by the general rate of interest. As interest rates rise, investors in stocks expect their yield to rise, which can only happen if the market price falls.

Shares are securities that provide title to the ownership of a company. Share holders of ordinary shares do not receive a fixed rate of interest but instead are entitled to a share in the company's profits. This is paid as a dividend. Shareholders are entitled to vote accompany meetings and upon liquidation of the company are entitled to a share of its assets

Store: (memory) A system or device used in computing to hold information (programs and data) in such a way that any piece of information can automatically be retrieved by the computer as required. The main store (or infernal store) of a computer is under the direct control of the central processor. If is the area in which prograins, or parts of programs, are stored white they are being run on the computer, Data and program instructions can be extracted extremely rapidly by random access. The main store is supplemented by backing store, in which information can be permanently stored. The two basic forms of backing store are those in which magnetic tape is used (i.e. magnetic tape units) and those in which disks, or sortie other random-access device, are used.

The main store is divided into a huge number of locations, each able to hold one unit of information, i.e., a word or a byte. The number of locations, i.e., the number of words or bytes that can be stored, gives the capacity of the store. Each location is identified by a serial number, known as its address.

There are many different ways in which memory can be classified. Random-access memory (RAM) and serial-access memory differ in the manner in which informa-tion is extracted from a store. With volatile memory, stored information is lost when the power supply is switched off, unlike nonvolatile memory, With read-only memory (ROM), information is stored permanently or semi-permanently; it cannot be altered by programmed instructions but can in some types be changed by special techniques.

Stores may be magnetic or electronic in character. Magnetic tape units, disk units, and the recently developed bubble memory are examples of magnetic memory. The electronic memory, widely used in main store consists of highly complex integrated circuits. This semiconductor memory stores an immense amount of information in a very small space, items of information can be extracted at very high speed.

STP: (NTP) Standard tem-perature and pressure, Conditions used internationally when measuring quantifies that vary with both pressure and temperature (such as the density of a gas). The values are 101325 Pa (approxi-mately 100 kPa) and 0°C (273.15 K)

Student's t-test: A hypothesis test for accepting or rejecting the hypothesis that the mean of a normal distribution with unknown variance is μ_0, using a small sample. The statistic $t = (x - \mu_0)\sqrt{n}/s$ is computed from the data ($x_1, x_2, \ldots x_n$), where x is the sample mean, s is the sample standard deviation, and $n < 30$. If the hypothesis, is true has a t,, distribution. If 1 lies in the critical region $|t| > I_{n-1}(I - \alpha/2)$ the hypothesis is rejected at significance level α.

subject: The main indepen-dent variable in an algebraic formula. For example, in the function

$$y\ f(x) = 2x^2 + 3x,$$

y is the subject of the formula.

Subnormal: The projection on the x-axis of a line normal to a curve at point $P_0(x_0, y_0)$ and extending front P to the x-axis. The length of the subnormal is my_0 (where m is the gradient of the tangent to the curve at P_0.

subroutine: A section of a computer program that performs a task that may be required several times in different parts of the program. Instead of inserting the same sequence of instruction at a number of different points, control is transferred to the subroutine and when the task is complete it is returned to the main part of the program.

substitution: A method of solving algebraic equations by replacing one variable by an equivalent in terms of another variable. For example, to solve the simultaneous equations

$$x + y = 4$$

and

$$2x + y = 9$$

we can first write x in terms of y, that is:

$$x = 4\ y$$

The substitution of 4y for x in the second equation gives:

$$2(4 - y) + y = 9$$

or $y = -1$, and therefore, from the first equation, $x = 5$. Another use of substitu-tion of variables is in integration.

subtangent: The projection on the x-axis of the fine tangent to a curve at a point P_o (x_o, y_o) and extending front P_0 to the x-axis. The length of the subtangent is y_0/m, where in is the gradient of the tangent.

Subtend: To lie opposite and mark out the limits of a fine or angle. For example, each arc of a circle subtends a particular angle at the centre of the circle.

Subtraction: Symbol The binary operation of finding the difference, between two quantities. In arithmetic, unlike addition, the subtraction of two numbers is neither commutative no associative. The identity element for arithmetic subtraction is zero only when it comes on the right hand side. In vector subtraction, two vectors are placed tail-to-tail forming two sides of a triangle. The length and direction of the third side gives the vector difference. Just as the sign of the difference between two numbers depends on the order of subtraction, the sense of the vector difference depends on the sense of the angle between the two vectors.

Superelastic collision: A collision for which the restitution coefficient is greater than one. In effect the relative velocity of the colliding objects after the interaction is greater than that before. The apparent energy gain is the result of transfer from energy within the colliding objects. For example, if a collision between two trolleys causes a compressed spring in one to be released against the other, the collision may be superelastic.

Supplementary units: The dimensionless units the radian and the steradian used along with base units to form derived units.

Surface: Any locus of points extending in two dimensions. It is defined as an area. A surface may he flat (a plane surface) or curved, finite or infinite. For example the plane $z = 0$ in three-dimensional Cartesian coordinates is flat and infinite; the outside of a sphere is curved and finite.

Surface of revolution: A surface generated by rotating a curve about an axis. For example, rotating a parabola about its axis of symmetry produces a paraboloid of revolution.

Symbolic logic: (formal logic) The branch of logic in which arguments, the terms used in them, the relationships between them, and the various operations that can be performed on them re all represented by symbols. The logical properties and implications of arguments can then be more easily studied strictly and formally, using algebraic techniques, proofs, and theorems in a mathematically rigorous way. It is sometimes called mathematical logic.

The simplest system of *symbolic logic* is preposi-tional logic (sometimes called prepositional calculus) in which letters, e.g. *P, Q, R,* etc., stand for propositions or statements, and various special symbols stand for relationships that can hold between them.

Symmetrical: Denoting any figure that can be divided into two parts that are mirror Images of each other. The letter A, for example, is symmetrical, and does not change when viewed in a mirror, but the letter *R* is not. A symmetrical plane figure has at least one fine that is an axis of symmetry, which divides it into two mirror images.

systems analysis: The detailed analysis of the activities of an organization or system; its basic purposes, and the needs that it must satisfy, so that its performance can be impro-ved or some other problem can be solved. The systems analyst reduces the steps required to improve a particular situation or solve a problem to a logical form. A suitable computer program can then be written to test or effect a solution, etc.

T

tangent: 1. A straight line or a plane that touches a curve or a surface without cutting through it. On a graph, the slope of the tangent to a curve is the slope of the curve at the point of contact. In Cartesian coordinates, the slope is the derivative dy/dx. If θ is the angle between the x-axis and a straight line to the point (x,y) from the origin, then the trigono-metrical function tanθ y/x.

2. (tan) A trigonometric function of an angle. The tangent of an angle α in a right-angled triangle is the ratio of the lengths of the side opposite to the side adjacent. This definition applies only to angles between 0° and 90° (0 and $\pi/2$ radians More generally, in rectangular Cartesian coordinates, with origin O, the ratio of the y-coordinates, to the x-coordinates of a point $P(x, y)$ is the tangent of the angle between the line OP and the x-axis. The tangent function like the sine cosine functions, is periodic but a repeats itself every 180° and is not continuous. Its zero when α = 0°, and becomes an infinitely large positive number. As a 90°, tan(–α) = –tanα, and so tanα approaches as α goes from 0° to –90°. At +90°, tanα jumps from to and there rises to zero at α = 180°.

tautology: In logic, a proposition statement, or sentence of a form that cannot post be false. For example 'if all pigs eat mice then some pigs eat mice' and 'if I am coming then I am coming' are both true regardless of the component propositions 'I am coming and all pigs eat mice' are true or false. More strictly, a tautology is a compound proposition that is true no matter what truth values are assigned to the simple propositions that it contains. A tautology is true purely because of the laws of logic and not because of any fact about the world. A tautology therefore contains no information.

tax: Money collected by a government from individuals or companies to fund its expenditure. *Direct taxes,* which are compulsory, include income tax on individual incomes, corpora-tion tax on the profits of companies, capital gains tax on additions in wealth, and capital transfer tax or gifts of money or property during a person's lifetime or after death (inheritance tax). *Indirect taxes* are levied on goods and services and are voluntary in the sense that they need only be paid if the goods or services taxed are purchased. They include taxes on petrol and value-added tax (VAT).

Taylor series: (Taylor expansion) A formula for expanding a function, $f(x)$ by writing it as an infinite series of derivatives for a fixed value of the variable, $x = a$:

$$f(x) = f(a)\ f'(a) + (x - a) +$$

$$f''(a)\ (x-a)^2/2+f''(x-a)^3\ 3! + \ldots$$

if $a = 0$, the formula becomes:

$$f(x) = f(0) + f(0)x + f'(0)\ x^1/2! + \ldots$$

This known as Maclaurin series, or Maclauria:: expansion.

tensor: A mathematical entity that is the general equivalent in any n-dimensional coordinate system, of vector in two or three-dimensional coordinates. Tensors are used to describe how all the components of quantity behave under certain transformations, just as a vector can describe a translation from one point to another in a plane or in space.

terminal: A point at which a user can communicate directly with a computer both for the input and output of information. It is situated outside the computer system, often at some distance from it, and is linked to it by electric cable, telephone, or some other transmission channel. A keyboard, similar to that on a typewriter, is used to feed information to the computer. The output can either be printed out, as with Teletype, or can be displayed unit. An interactive terminal is one connected to a computer which give an almost immediately response to an enquiry from the user. An intelligent terminal can store information and perform simple operations on it without the assistance of the computers central processor.

time sharing: A method of operation in computer systems in which a number of jobs are apparently executed simultaneously instead of one after another (as in batch processing). This is achieved by transferring circle program in turn from backing store to main store and allowing it run for a short time. Timesharing is particularly useful for programs that are controlled by users at terminals. It allows them all to interact with the computer apparently simultaneously, provided there are not too many of them.

ton: 1. A unit of mass equal to 2240 pounds. It is equivalent to 1016.05 kg 2. A unit used to express the explosive power of a nuclear weapon. It is equal to an explosion with an energy equivalent of one tone of TNT or approximately 5×10^9 joules.

tonne: (metric ton) Symbol A unit of mass equal to 10^3 kilograms.

topological space: A set X that has a set T of all its subsets, satisfying the conditions: $\phi \in T$; $X \in T$ and if $U \in T$ and $V \in T$ then ($(Y \cap V)$; $\in T$ and $(U \cap V) \in T$. The members of T are called the open set of the topological space X. Any set of Points forming a geometrical figure and satisfying these conditions is a topological space. Its topology is defined by the properties of its open sets.

topology: A branch of geometry concerned with the general properties of shapes and space. It can be thought of as the study of properties that are not changed by continuous deformations, socle as stretching or twisting. A sphere and an ellipsoid are different figures in solid (Euclidean) geometry, but in topology they are considered equivalent since one can be transformed into the other by a continuous deformation. A torus, on the other hand is not topologically equivalent it to a sphere–it would not be possible to distort a sphere into a torus without breaking or joining surfaces. A torus is thus a different type of shape from a sphere. Topology studies types of shapes and their properties. A special case of this is the investigation of networks of lines and the properties of knots.

In fact Euler's study of the Konisberg bridge problem was one of the earliest results in topŏlogy. A modern example is in the analysis of electrical circuits. A circuit diagram is not an exact reproduction of the paths of the wires, but does show the connections between diffe-rent points of the circuit it is topologically equiva-lent). In printed or integrated circuits if is important to arrange connections so that they do not cross.

Topology uses methods of higher algebra including group theory and set theory. An important notion is fruit of sets of points, and of points in the neighborhood of a given point (i.e., within a certain distance of the point). An open set is a set of points such that each point in the set has a neighborhood containing points in the set. A topŏlogical transformation occurs when there is a one-to-one correspondence between points in one figure and points in another so that open sets in one correspond to open sets in the other. If one figure can be transformed into another by such a transformation, the sets are topologically equivalent.

torque: Symbol T A turning force (or moment). The torque of a force Fs about an axis (or point) is a where s is the distance from the axis to the fine of action of the force. The unit is the newton metre. Note that the unit of work, also the newton metre, is called the joule. Torque is not however measured in joules. The two physical quantities are not in lac, the same. Work (a scalar) is the scalar product of force and displacement. Torque is the vector product and is a vector at 90° to the plane of the force and displacement.

torr: A unit of pressure equal to a pressure of 101 325/760 pascals (133.322 Pa). It is equal to the mmHg.

torsional wave: A wave motion in which the vibrations in the medium are rotatory simple harmonic motions around the direction of energy transfer.

torus: (anchor ring) A closed curved surface with a hole in it, like a tyre inner tube. If can be generated by rotating a circle about an axis that lies in the same plane as the circle but does not cut if. A cross-section of the torus in a plane perpendicular to the axis is two concentric circles. A cross-section in any plane that contains the axis is a pair of congruent circles at equal distances on both sides of the axis. The volume of a tortus is $4\pi dr^2$ and its surface area is $3\pi^2 dr$, where r is the radius of the generating circle and d is the distance of its centre from the axis. In Cartesian coordinates, a torus that has its axes lying along the z-axis and a generating circle in the y-z plane with its centre at a distance d along the y-axis, has the equation:

$$\sqrt{[(x^2 + y^2) - d^2]} + z^2 = r^2$$

total derivative: A derivative that can be expressed in terms of a series of partial derivatives. For example, if the function $z = f(x, y)$ is a continuous direction of x and y, and both x and y are continuous functions of another variable t, then the total derivative of z with respect to t is:

$$dz/dt \ (\partial z/\partial x)(dx/dt) + (\partial z/\partial y)(dy/dt).$$

total differential: An infinitesimal change in a function of one or more variables. It is the sum of all the partial differentials.

transformation: 1. In general, any function or mapping that change one quantity into another. The changing of an algebraic expression or equation into an equivalent with a different form. For example, the equation

$$(x-3)^2 = 4x + 2$$

can be transformed into

$$x^2 - 10x - 11 = 0.$$

In geometry, the changing of one shape into another by moving each point in it to, a different position, usually by a specified procedure. For example, a plane figure may be moved in relation to two rectangular axes. Another example is when a figure is enlarged.

transformation of coordinates: 1. Changing the position of the reference axis in a coordinate system by translation, rotation, or both, usually to simplify the equation of a curve. 2. Changing the type of coordinate system in which a geometrical figure is described. For example, from rectangular coordinates to polar coordinates.

translation: The moving of a geometrical figure so that only its position relative to fixed axes is changed, but not its orientation, size or shape.

translation of axes: In coordinate geometry, the shifting of the reference axes so that each axis is parallel to its original position and each point is given a new set of coordinates. For example, the origin O of a system of x– and y-axes, may be shifted to the point. $O'(3, 2,)$ in the original system. The new axes x' and y' are at $x = 3$ and $y = 2$, respectively. This is sometimes done to simplify the equation of a curve. The circle $(x - 3)^2 + (y - 2)^2 = 4$ can be described by new coordinates $x' = (x - 3)$ and $y' = (y - 3)$: $(x')^2 + (y')^2 = 4$. The origin O' is then at the centre of the circle.

translatory motion: (transla-tion) Motion involving change of position; it compares with rotatory motion (rotation) and vibratory motion (vibration). Each is associated with kinetic energy. In an object undergoing translatory motion, all the points move in parallel paths. Translatory motion is usually described in terms of (linear) speed or velocity, and acceleration.

transpose of a matrix: The matrix that results front interchanging the rows and columns of a matrix. The determinant of the transpose of a square matrix is equal to that of the original matrix. The transpose of a row vector is a column vector and vice versa. If two matrices A and B are conformable (can be multiplied together, then the transpose of the matrix product $AB = C$ is = $(AB) = BA$. In other words, in taking the transpose of a matrix product the order must be reversed.

Transverse wave: A wave motion in which the motion or change in perpendicular to the direction of energy transfer. Electromagnetic waves and water waves are examples of transverse waves.

trapezium: A plane figure with four straight, sides, two of which are parallel. Its area is the product of the sum of the parallel side lengths and hall the perpendicular distance between them.

trapezium rule: A rule for finding the approximate area under a curve by dividing it into pairs of trapezium shaped sections, forming vertical columns of equal width with bases lying on the horizontal axis. The trapezium rule is used as a method of numerical integration. For example, if the value of a function $f(x)$ is known at $x = a$, $x = b$, and at a value midway between a and b, the integral is approximately:

$$(h/2)[f(a)+2f((a+b)/2)+f(b)]$$

h is half the distance between a and b. This does not give a sufficiently accurate result, the area may be subdivided into 4, 6, 8, ... columns until further subdivision makes no significant difference to the result.

triangle: A plane figure with three straight sides. The area of a triangle is a hall the length of one side, the base, times the length of the perpendicular line, the altitude, front the base to the opposite vertex. The sum of the internal angles in a triangle is 180° (or π radians). In an *equilateral* triangle, all three sides have the same length and all three angles are 60°. An *isosceles triangle* has two sides of equal length and two equal angles. A *scalene triangle* has no equal sides or angles. In a right-angled triangle, one angle is 90° (or $\pi/2$ radians), and the others therefore add up to 90°. In an acute *angled triangle*, all the angles are less than 90°. In an *obtuse-angled triangle*, one of the angles is greater than 90°.

triangle inequality: For any triangle ABC, the length of one side is always less than the sum of the other two:

$$AB < BC + CA$$

triangle of vectors: A triangle describing three coplanar vectors acting at a point with zero resultant. When drawn to scale shown correctly in size, direction, and sense, but not in position they form a closed triangle. Thus three forces acting on an object ai equilibrium form a triangle of forces. Similarly a triangle of velocities can be constructed.

Triangular matrix: A square matrix in which either ail the elements above the leading diagonal or ail the elements below the leading diagonal are zero. The determinant of a trilingual matrix is the product of its diagonal elements.

Triangular numbers: The set of numbers 1, 3, 6, 10,...) generated by triangular arrays of dots. Each array has one more row than the preceding one, the additional row having one more dot than the longest in the preceding array. The *n*th triangular number is $n(n+1)/2$.

trigonometry: The study of the relationships between the sides and angle in triangle, in terms of the *trigonometric functions* of angles (sine, cosine, and tangent). *Trigonometric functions*, can be defined by the ratio of sides in a right-angled triangle. In a right-angled triangle with an angle a, o is the length of the side opposite α, a the length of side adjacent to, α, and h the length of the hpotenuse. For such a triangle the three trigonometric functions of α are defined by:

$$\sin\alpha = o/h$$

$$\cos\alpha = a/h$$

$$\tan\alpha = o/a$$

The following relationships hold for all values of the variable angle α:

$$\cos\alpha = \sin(a + 90°)$$

$$\cos^2\alpha = + \sin^2\alpha = 1$$

$$\tan\alpha = \sin^2\alpha/c\,os\alpha$$

The trigonometric functions of an angle can also be defined in terms of a circle (hence they are sometimes called circular functions). A circle is taken with its centre ai the origin of Cartesian coordinates. If a point P is taken on this circle, the fine OP makes an angle α with the positive direction of the *x*-axis.

Then, the trigonometric functions are:

$$\tan\alpha = \psi/x$$

$$\sin\alpha = y/\text{OP}$$

$$\cos\alpha = x/\text{Op}$$

Here, (x,y) are the coordinates of the point P and OP = $\sqrt{(\xi^2 + y^2)}$ The signs of x and y are taken into account. For example, for an angle β° between 90° and 180°, y will be positive and x negative. Then:

$$\tan\beta = -\tan(180 - \beta)$$

$$\sin\beta = + \sin(180 - \beta)$$

$$\cos\beta = -\cos(180 - \beta)$$

Similar relationships can be written for the trigonometric functions of angles between 180° and 270° and between 270° and 360°.

The functions secant (sec), cosecant (cosec), and cotangent (cot), which are the reciprocals of the cosine, sine, and tangent functions respectively, obey the following rules for all values of α:

$$\tan^2\alpha + 1 = \sec^2\alpha$$

$$1 + \cos^2\alpha = \operatorname{cosec}^2\alpha$$

trinomial: An algebraic expression with three variables in it. For example, $2x + 2y + z$ and $3a + b = c$ are trinomials.

triple integral: The result of integrating the same function three times For example, if a function $f(x,y,z)$ is integrated first with respect to x, holding y and z constant, and the result is then integrated with respect to y, holding x and z constant, and the resultant double integral is then integrated with respect to z holding x and y constant, the triple integral is

$$\iiint f(x, y, z)dzdydx.$$

Triple scalar product: A product of three vectors, the result of which is as scalar, defined as:

$$A.(B \times C = ABC = \text{ABC} \sin\theta \cos\phi$$

where θ is the angle between A and the vector product $(B \times C)$,θ being the angle between B and C. The scalar triple product is equal to the volume of the parallelepiped of which A, B, and C form nonparallel edges. If A, B, and Care coplanar, their triple scalar product is zero.

triple vector product: A product of three vectors, the result of which is a vector. It is a vector product of two vectors, one of which is itself a vector product. That is:

$$A\times(B\times C) = (A.C)B - (A.B)C$$

Similarly

$$(A\times B)\times C = (A.C)B - (B\times C)A.$$

These are equal only when A, B, and C are mutually perpendicular.

Trivial solution: A solution to an equation or set of equations that is obvious and gives no useful information about the relationships between the variables involved. For example, $x^2 + y^2 = 2x + 4y$ has a trivial solution $x = 0$; $y = 0$

truncated: Describing a solid generated from a given solid by two nonparallel plane cutting the given solid.

truth table: In logic, a mechanical procedure (some-times called a matrix) that can be used to define certain logical operations, and to find the truth value of complex propositions or statements containing combinations of simpler ones. A truth table lists, in rows, all the possible combinations of truth values (T ='true', F ='false') of a proposition or statement, and given an initial assignment of truth or falsity to the constituent parts, mechanically assigns a value to the whole. The truth table definitions for conjunction, disjunction, negation and implication are shown at those headwords.

An example of a truth table for a complex, or compound proposition is shown in the illustration. The assigning of values proceeds in this way: on the basis of the truth values of P and Q, the simple propositions are given truth values, written under the signs (Λ in $P \Lambda Q$, ~ in ~P). Using these, truth values can then be assigned to the whole; in the example this is in effect a complex disjunction and the values are written under the sign.

Thus in the case where P is true and Q is false, $P \Lambda Q$ will be false, P will be false, and therefore, the whole will be false.

truth value: The truth or falsity of a proposition is logic. A true statement or proposition is indicated by T and a false one by F. In computer logic, the digits 1 and 0 are often used to denote truth value.

Turning point: A point on the graph of a function at which the slope (gradient) of the tangent to a continuous curve changes its sign. If the slope changes from positive to negative, that is the coordinate stops increasing and starts decreasing, it is a maximum point. If the slope changes from negative to positive it is a minimum point. Turning points may be local maxima and minima or absolute maxima and minima. All turning points are stationary points. At a turning point the derivative, dy/dx, of the curve $y = f(x)$ is zero.

Two dimensional: Having length and breadth but not depth. Flat shapes, such as circles, squares, and ellipses, can be described in a coördinate system using only two variables, for example, two dimensional Cartesian coordinates with an x-axis and a y-axis.

U

uniform motion A vague phrase, usually taken to Mean motion at constant velocity, often in a straight line.

uniform velocity Constant velocity, describing motion in a straight fine with zero acceleration,

union Symbol ∪ The combined set of all the element of two or more sets. If A = {2, 4, 6} and B = {3, 6, 9} then A∪B (2, 3, 4, 6, 9).

unique solution The only possible value of a variable can satisfy an equation. For example, $x + 2 = 4$ has the unique solution $x = 2$, but $x = 4$ has unique solution because $x = + 2$ and $x = -2$ both satisfy the equation.

unit A reference value of a quantity used to express other values of the same quantity.

unit matrix (identity matrix) symbol I A square matrix in which the element in the leading diagonal are all equal to one and the other elements are zero if a matrix A with m rows and n column, is multiplied by an $n \times n$ unit matrix, I, it remains unchanged that is $IA = A$. The unit matrix is the identity matrix for matrix multiplication.

unit vector A vector with a magnitude of one unit. Any vector can be expressed in terms of magnitude the scalar quantity and the unit vector r¢ which has the direction as $r.r. = rr'$. In three dimen-sional Cartesian coordinates with origin 0 unit vectors i, j and k are used in the x–, y–, and z-directions respectively.

universal set Symbol : E or ε. The set that contains all possible elements. In a particular problem, E will be defined according to the scope of the Problem. For example, in a calculation involving only positive numbers the universal ⁺set, E, is the set of all positive numbers.

unstable equilibrium Equili-brium such that if the system is disturbed a little, there is a tendency for it to move further from its original position rather than to return.

upthrust An upward force on an object in a fluid. In a fluid in a gravitational field the pressure increases with depth. The pressures at different points on the object will therefore differ and the resultant is vertically up-ward.

utility programs Programs that help in the general running of a computer system. They can be used, for example, to make copies of files and to transfer data from one storage device to another, as from a magnetic tape unit to a disk store.

V

validity: in logic, a property of arguments, inferences, or deductions. An argument is valid if it is impossible that the conclusion be false while the premisses are true. That is to assert the premisses and deny the conclusion would be a contradiction.

variable: A changing quantity, usually denoted by a letter in algebraic equations, that might have any one of a range of possible values. Calculations can be carried out on variables because certain rules apply to all the possible values. For example to carry out the operation of squaring all the integers between 0 and 10, an equation can be written in terms of an integer variable n: $y = n^2$ with the condition that is between 0 and 10 ($0 < n < 10$). y is called a dependent variable because its value depends on the value chosen for n i.e. it can only have the values 1, 4, 9.. etc. An independent variable has no such relationship with another variable. For example, if one variable, x denotes the number of students in a school and another, y denotes the proportion of the total number of students who want in take school lunches, then x and y are independent variables and a change in either of them will not affect the other. However, their product, xy, will affect a third quantity the number of lunches ordered. Variables may also denote quantifies other than ordinary arithmetic numbers, for example, vector variables and matrix variables.

variance: A measure of the dispersion of a statistical sample. In a sample of n observations $x_1, x_2, x_3, \ldots x_n$ with a sample mean x the sample variance is $s^2 = [(x_1 - x]^2 + (x_2 - x)^2 + (x_3 - x)^2 + (x_n - x)^2{}_]$ / $n - 1$.

vector: A measure in which direction is important and must usually be specified. For instance, displacement is a vector quantity, whereas distance is a scalar, weight, velocity and magnetic field strength are other examples of vectors they are each quoted as a number with a unit and a direction. Vectors are denoted by printing the in bold italic type F. Vector algebra treats vectors symbolically in a similar way to the algebra of scalar quantities but with, different rules for addition, subtraction, multiplication, etc.

Any vector can be represented in terms of component vectors. In particular a vector in three dimensional Cartesian coordinates can be represen-ted in terms of three unit vector components *i*, *j* and *k* is directed along the *x*-, *y*-, and z-axes respectively. If P is a point with coordinates (x_1, y, z_1) then the vector OP is equivalent to $ixi, + jy_1 + kz_1$

vector difference: The result of subtracting to vectors. On a vector diagram, two vectors, A and B, are subtracted by drawing them tail-to-tail. The difference A – B is the vector represented by the line from the head of B to the head of A. If A and B are parallel, the magnitude of the difference is the difference of the individual magnitudes. If they are antiparallel, it is the sum of the individual magnitudes. The vector difference may also be calculated by taking the difference in the magnitudes of the corresponding components of each vector. For example, for two plane vectors in a Cartesian coordinates system.

$$A = 4i + 2j$$

$$B = 2i + j$$

where *i* and *j* are the unit vectors parallel to the *x* and *y* axes respectively,

$$A - B = 2i - j$$

vector multiplication: Multi-plication of two or more vectors. This can be defined in two ways according to whether the result is a vector or a scalar.

vector product: A multi-plication of two vectors to give a vector. The vector product of A and B is written A × B is a vector of magnitude AB sinθ where A and B are the magnitudes of A and B and θ is the angle between A and B The direction of vector product is at right angles to A and B. It points in the direction in which a right-handed screw would move turning from A towards B. An example of a vector product is force F on a moving charge Q in a field B with velocity v (as in the motion effect). Here

$$F = QB \times v$$

Another example is the product of a force and a distance to give a moment (turning effect), which can be represented a vector at right angles to the plane in which the turning effect acts. The vector product is sometimes called the cross product. The vector product is not commutative because

$$A \times B = -(B \times A)$$

It is distributive with respect to vector addition

$$C\times(A\times B) = (C\times A) + (C\times B)$$

The magnitude of A × B is equal to the area of the parallelogram of which A and B form two nonparallel sides, three-dimensional Cartesian coordinate system with unit vectors *i*, *j*, and *k* in the *x*, *y*, and *z*, directions respectively,

$$A \times B = (a_1 i + a_2 j + a_3 k) \times$$
$$(b_1 i + b_2 j + b_3 k)$$

This can also be written as a determinant.

vector projection: The vector resulting from an orthogonal projection of one vector on another. For example, the vector projection of *A* on *B* is bAcosθ where θ is the smaller angle between A and B, and b is the unit vector in the direction of B.

vector sum: The result of adding two vectors. On a vector diagram, vectors are added by drawing them head to tail. The sum is the vector represented by the straight line from the tail of the first to the head of the last. If they are parallel, the magnitude of the sum is the sum of the magnitudes if the individual vectors. If two vectors are antiparallel then the magnitude of the sum of the difference of individual magnitudes.

The vector sum may also be calculated by summing the magnitudes of the corresponding components of each individual vector. For example, for two plane vectors, $A = 2i + 3j$ and $B = 6i + 4j$, in a Cartesian coordinate system with unit vectors *i* and *j* parallel to the x- and y-axes respectively, the vector sum, A + B, equals $8i + 4j$.

velocity: Symbol: v Displacement per unit time. The unit is the metre per second (ms^{-1}). Velocity is a vector quantity, speed being the scalar form. If velocity is constant, it is given by the slope of a position/time graph, and by the displacement divided by the time taken. If it is net constant, the mean value is obtained. If *x* is the displacement, the instanta-neous velocity is given by

$$v = dx/dt$$

Venn diagram: A diagram used to show the relationships between sets. The universal set, *E*, is shown as a rectangle. Inside this other sets are shown as circles. Intersecting or overlapping circles are intersecting sets. Separate circles are sets that have no intersection. A circle inside another is a subset. A group of elements or subset, defined by any of these relationships may be indicated by a shaded area in the diagram.

vertex: 1. A point at which lines or planes meet in a figure; for example, the top point of a cone or pyramid, or a corner of a polygon or polyhedron. 2. One of the two points to which an axis of a cone cuts the conic.

vertically opposite angles: One of the two pairs of equal angles formed when two straight fines cross each other.

vibration: (oscillation) Any regularly repeated to-and-fro motion or change. Examples are the swing of a pendulum, the vibration of a sound source, and the change with lime of the electric and magnetic fields in an electromagnetic wave.

virtual work: The work done if a system is displaced infinitesimally from its position. The virtual work is zero if the system is in equilibrium.

visual display unit (VDU): A computer terminal with which the user can communicate with the computer by means of a keyboard similar to that or a typewriter; this input and also the output from the computer appears on a television screen. A VDU can operate both as an input and output device. The information displayed is in the form of words, numbers, etc., A *graphics display* is a similar device in which information appears as graphs or other drawings as well as text.

volt: Symbol V The SI unit of electrical potential, potential difference, and e.m.f. defined as the potential difference between two points in a circuit between which a con-stant current of one ampere flows when the power dissi-pated is one watt. One volt is one joule per coulomb.

volume: Symbol: *V* The extent of the space occupied by a solid or bounded by a closed surface, measured in units of length cubed. The volume of a box is the product of its length, its breadth, and its height. The SI unit of volume is the cubic metre (m^3).

W

wave: A method of energy transfer involving some form of vibration. For instance, waves on the surface of a liquid or along a stretched string involve regular to-and-fro motion of particle about a mean position. Sound waves carry energy by alternate compressions and rarefactions of air (or other media) in electromagnetic waves, electric and magnetic fields vary at right angles to, the direction of propagation of the wave. At any particular instance, a graph of displacement against distance is a regular repeating curve the wave form or wave profile of the wave in a travelling (or progressive) Wave the whole periodic displacements moves through the medium. At any point in the medium the disturbance is changing with lime under certain conditions a stationary (or standing) wave can be produced in which the disturbance does not change with time.

For the simple case of a plane progressive wave the displacement at a point can be represented by an equation:

$$y = a\sin 2\pi(ft - x/\lambda)$$

where a is the amplitude, f the frequency, x the distance from the origin, and λ the wavelength. Other relation-ships are:

$$y = a\sin 2\tau(vt - x)/\lambda$$

where v is the speed, and

$$y = a\sin 2\pi(t/T - x/\lambda)$$

where T is the period. Note that if the – sign is replaced by a + sign in the above equations it implies a similar wave moving in the opposite direction. For a stationary wave resulting from two waves in opposite directions, the displacement is given by:

$$y = 2a\cos 2\pi x/\lambda$$

wave equation: A second order partial differential equation that describes wave motion. The equation

$$\partial^2 u / \partial x^2 = (1/c^2)\, \partial^2 u / \partial t^2$$

might represent, for example, the vertical displacement u of the water surface as a plane wave of velocity c passes along tire surface, the horizontal position and the time being given by x and t respectively The general solution of this one-dimensional wave equation is a periodic function of x and t.

wavefront: A continuous surface associated with a wave radiation, in which all the vibrations concerned are in phase. A parallel beam has plane wavefronts; the output of a point source has spherical wavefronts.

wavelength: Symbol: The distance between the ends of one complete cycle of a wave. Wavelength relates to the wave speed (c) and its frequency (v) thus:

$$c = v\lambda$$

wave motion: Any form of energy transfer that may be described as q wave rather than as a stream of particles. The term is also sometimes used to mean any harmonic motion.

wave number: Symbol: σ The reciprocal of the wavelength of a wave. It is the number of wave cycles in unit distance, and is often used in spectroscopy. The unit is the metre^{-1} (m^{-1}). The circular wave number (Symbol: k) is given by

$$k = 2\pi\sigma.$$

weber: Symbol: Wb The SI unit of magnetic flux, equal to the magnetic flux that, linking a circuit of one turn, produces art e.m.f, of one volt when reduced to zero at a uniform rate in one second. 1 Wb = 1Vs.

weight: Symbol: W The force by which a mass is attracted to another, such as the Earth. It is proportional to the body's mass (m), the constant of proportionality being the gravitational field strength (i.e. the acceleration of free fall). Thus W = mg where g is the acceleration of free fall. The mass of a body is normally constant, but its weight varies with position (because it depends on g).

weightlessness: An apparent loss of weight experienced by an object in free fall. Thus for a person in an orbiting spacecraft, the weight in the Earth's frame of reference is the centripetal force necessary to maintain the circular orbit. In the frame of reference of the spacecraft, the person feels that he has no weight.

wheel and axle: A simple machine consisting of a wheel on an axle that has a rope around it. An effort applied to the wheel is transmitted to a load exerted at the axle tope. The force ratio (mechanical advantage) is equal to r_W/r_A where r_W is the radius of the wheel and r_A that of the axle.

word: The basic unit in which information is stored and manipulated in a computer. Each word consists usually of a fixed number of bits. This number is known as the word length.

work: Symbol: W The work done by a force is the product of the force and the distance moved in the same direction.

$$\text{work} = \text{force} \times \text{displacement}$$

work is in fact a process of energy transfer and, like energy, is measured in joules. If the directions of force (F) and motion are not the same the component of the force in the direction of the motion is used.

$$W = F\cos\theta$$

where s is displacement and θ the angle between the directions of force and motion. Work is the scalar product force and displace-ment.

Y

yield: The income produced by a stock or share expressed a percentage of its market price. For example a share with a nominal value of 50p pays a 12% dividend it will pay 6p per share. If the market price is 80p, the yield would be (6' 100)/80 = 75%. In the case of stocks that pay fixed interest, the yield to a large extent determines the market price. The buyers of stocks will expect a yield comparable to the current interest if these are rising the yeild of fixed interest stocks must also rise. This can only happen if the market price falls. The rising interest rates depress the market price of stock and vice versa.

Z

zero: (0) The number that when added to another number gives a sum equal to that the other number. It is included in the set of integers but not in the set of whole numbers. The product of any number and zero is zero. Zero is the identity element of addition.

zone: A part of a sphere produced by two parallel planes cutting the sphere.